A Manual of Engineering Drawing Practice

Colin H. Simmons
Mem ASME, FIED

Standards Engineer, CAV Ltd, A Lucas Company

Dennis E. Maguire
CEng, MIMechE, Mem ASME, MIED

Senior Lecturer, Mechanical and Production Engineering Department, Southall College of Technology

HODDER AND STOUGHTON
LONDON SYDNEY AUCKLAND TORONTO

ISBN 0 340 17996 1

First published 1974
Sixth impression 1985

620.004 S

Typesetting by Print Origination, Liverpool, England
Printed in Great Britain for
Hodder and Stoughton Educational.
a division of Hodder and Stoughton Limited,
Mill Road, Dunton Green, Sevenoaks, Kent,
by Hazell Watson & Viney Ltd

Contents

Preface

1 The drawing office   1
2 Drawing instruments and materials   3
3 Linework and lettering   6
4 Microfilming   10
5 Scales   12
6 Practical geometry   14
7 Tangency   17
8 Loci   20
9 First- and third-angle projection   30
10 Isometric and oblique projection   33
11 Reading engineering drawings   38
12 Conic sections   39
13 True lengths and auxiliary views   42
14 Interpenetration   47
15 Development   51
16 Types of drawings and layouts   59
17 Simplified drawings   64
18 Sections   65
19 Dimensioning   68
20 Machining and surface roughness   71
21 Screw-threads   74
22 Conventional representation of screw-threads   77
23 Nuts, bolts, screws, and washers   78
24 Counterbores, countersinks, and spotfaces   83
25 Tapers and chamfers   85
26 Keys and keyways   88
27 Springs   92
28 Welding and welding symbols   99
29 Limits and fits   103
30 Geometrical tolerances   108
31 Datums   111
32 Application of geometrical tolerances   113
33 Maximum material condition   125
34 Co-ordinate and positional tolerances   131
35 Cams   136
36 Gears   142
37 Ball and roller bearings   149
38 Adhesive bonding   162
39 Production drawings   167

Preface

This book is intended to provide a general guide to engineering draughtsmanship, and we hope that the contents will equally assist all engineering students and practising draughtsmen. It has not been written with any one particular examination course in mind, since we feel that this material is required, in full or in part, by all students who study the subjects of Engineering Drawing and Engineering Design on TEC, City and Guilds, 'O' and 'A' level, National Certificate and Degree Courses.

Engineering drawing is an international language used to convey impressions and instructions from a designer to the user, inspector, or manufacturer. Drawings should be produced in accordance with the current British Standards and International Standards Organisation (ISO) recommendations, and our intention in writing this book is also to show how these standards are interpreted by draughtsmen on simple engineering applications.

Due to lack of space we have, with regret, been unable to include student exercises, but the worked examples in each section have been carefully selected to show certain draughting principles relating to that particular aspect of engineering. With this fact in mind, any of the geometrical types of drawing could well be copied for course work, after estimating leading dimensions. Sizes are relatively unimportant for an understanding of the working principles involved.

We also hope to have made a contribution towards a better understanding of the very important engineering topic of geometrical tolerancing. Considerable space has been allocated to this subject, which, perhaps because it is relatively new, is not widely practised by smaller companies. Many examples are provided, to show how each type of tolerance can be applied.

We are most grateful to the British Standards Institution for permission to reprint extracts from their publications, also to the companies listed below for their kind assistance with the reproduction of information and illustrations. Thanks too for the encouragement, constructive suggestions, and time spent checking drawings and scripts by our colleagues. Our final thanks go to our patient and understanding wives, Audrey and Beryl, who gave us much clerical assistance after we had just failed to recruit them as apprentice draughtswomen.

May we sincerely wish all students and draughtsmen who use this volume every success in their studies and careers.

C. H. Simmons
D. E. Maguire

Acknowledgements

The authors express their special thanks to: Airfix Products Ltd, London SW 18; Barber and Colman Ltd, Sale, Cheshire; British Standards Institution, 2 Park Street, London W1A 2BS, from whom complete copies of the standards quoted may be obtained; CIBA-GEIGY (UK) Ltd, Duxford, Cambridge; Firth Brown Ltd, Atlas Works, Sheffield; Lepton Engineering, 147 High Street, Chalfont St Peter, Bucks; Ransome Hoffman Pollard Ltd, Chelmsford, Essex; F.S. Ratcliffe (Rochdale) Ltd, Crawford Spring Works, Rochdale; and Staedler (UK) Ltd, Pontyclun, Glamorgan.

CHS
DEM

Chapter 1

The drawing office

The drawing office is generally regarded as the heart of any manufacturing organisation. Products, components, ideas, layouts, or schemes which may be presented by a designer in the form of rough freehand sketches, or developed stage by stage on the board, are translated into working drawings by the draughtsman. There is generally very little constructive work that can be done by other departments within the firm without a drawing of some form being available. The drawing is accepted as the universal means of communication, and BS 308: 1972 is the current standard for engineering drawing practice.

The British Standard yearbook lists all the current standards, and the draughtsman should ensure that his complete drawing conforms to BS 308 regarding presentation, and also that the contents of the drawing are themselves, where applicable, in agreement with separate standards relating to materials, dimensions, processes, etc. Larger organisations employ standards engineers who ensure that products conform to British and also international standards where necessary. Good design is often the product of teamwork where detailed consideration is given to the aesthetic, economic, ergonomic, and technical aspects of a given problem. It is therefore necessary to impose the appropriate standards at the design stage, since all manufacturing instructions originate from this point.

A perfect drawing communicates an exact requirement, or specification, which cannot be misinterpreted and which can form part of a legal contract between supplier and user.

Engineering drawings can be produced to a good professional standard if the following points are observed:
a) linework must be of uniform thickness and density;
b) always use instruments—never mix freehand sketching with machine drawing;
c) eliminate fancy printing, shading, and associated artistry;
d) include on the drawing only the information which is required to ensure accurate clear communication;
e) where possible, use only standard symbols and abbreviations;
f) use draughting aids and templates for repetitive work;
g) ensure that the drawing is correctly dimensioned (adequately but not over-dimensioned) with no unnecessary details.

Remember that care and consideration given to small details makes a big contribution towards perfection, but that perfection itself is no small thing. An accurate, well delineated engineering drawing can give the draughtsman responsible considerable pride and job satisfaction.

The field of activity of the draughtsman may involve the use, or an appreciation, of the following topics.

1 *Company communications* Most companies have their own systems which have been developed over a period of time for the following:
   a) internal paperwork,
   b) numbering of drawings and contracts,
   c) coding of parts and assemblies,
   d) production planning for component manufacture,
   e) quality control and inspection,
   f) updating, modification, and reissuing of drawings.

2 *Company standards* Many drawing offices use their own standard methods which arise from satisfactory past experience of a particular product or process. Also, particular styles may be retained for easy identification, e.g. certain prestige cars can be easily recognised since some individual details, in principle, are common to all models.

3 *Standards for dimensioning* Interchangeability and quality are controlled by the application of practical limits, fits, and geometrical tolerances.

4 *Material standards* Physical and chemical properties, non-destructive testing. Preferred sizes, stock sizes, and availability of rod, bar, tube, plate, sheet, nuts, bolts, rivets, etc. and other bought-out items.

5 *Draughting standards and codes of practice* Drawings must conform to accepted standards, but components are sometimes required which in addition must conform to certain local requirements or specific regulations, for example relating to safety when operating in certain environments or conditions. Assemblies may be required to be flameproof, gastight, waterproof, or resistant to corrosive attack, and detailed specifications from the user may be applicable.

6 *Standard parts* are sometimes manufactured in quantity by a company, and are used in several different assemblies. The use of standard parts sometimes reduces an unnecessary variety of materials and basically similar components.

7 *Standards for costs* The draughtsman often requires to compare costs where different methods of manufacture are available. A component could possibly be made by forging, by casting, or by fabricating and welding, and a decision as to which method to use must be made. The draughtsman must obviously be well aware of the manufacturing facilities and capacity offered by his own company, the costs involved when different techniques of production are employed, and also an indication of the costs likely when work is sub-contracted to specialist manufacturers, since this alternative often proves an economic proposition.

8 *Data sheets* Tables of sizes, performance graphs, and conversion charts are of considerable assistance to the design draughtsman.

Fig. 1.1 shows the main sources of work flowing into a typical industrial drawing office. The drawing office provides a service to each of these sources of supply, and the work involved can be classified as follows.

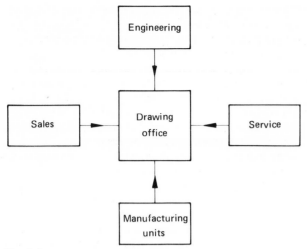

Fig. 1.1

1 *Engineering* The engineering departments are engaged on
   a) current production;
   b) development;
   c) research;
   d) manufacturing techniques, which may include a study
      of metallurgy, heat-treatment, strength of materials,
      and manufacturing processes;
   e) advanced project planning;
   f) field testing of products.
2 *Sales* This department covers all aspects of marketing
   existing products and market research for future products. The drawing office can receive work in connection with
   a) general arrangement and outline drawings for pros-
      pective customers;
   b) illustrations, charts, and graphs for technical publica-
      tions;
   c) modifications to production units to suit customers'
      particular requirements;
   d) application and installation diagrams;
   e) feasability investigations.
3 *Service* The service department provides a reliable,
   prompt, and efficient after-sales service to the customer.
   The drawing office receives work associated with
   a) maintenance tools and equipment;
   b) service kits for overhauls;
   c) modifications to production parts resulting from field
      experience;
   d) service manuals
4 *Manufacturing units* Briefly, these cover all departments
   involved in producing the finished end-product. The
   drawing office must supply charts, drawings, schedules,
   etc. as follows:
   a) working drawings of all the company's products;

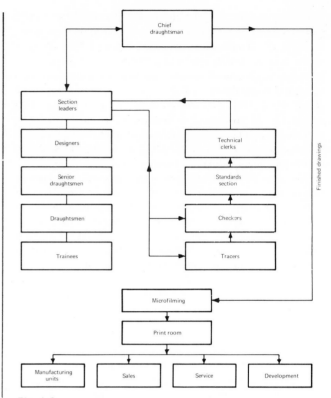

Fig. 1.2

   b) drawings of jigs and fixtures associated with manufac-
      ture;
   c) plant-layout and maintenance drawings;
   d) modification drawings required to aid production;
   e) reissued drawings for updated equipment;
   f) drawings resulting from value analysis and works'
      suggestions.

Fig. 1.2 shows the organisation in a typical drawing
office. The function of the chief draughtsman is to take
overall control of the services provided by the office. The
chief draughtsman receives all work coming into the
drawing office, which he examines and distributes to the
appropriate section leader. The section leader is responsible
for a team of draughtsmen of various grades. When work is
completed, the section leader then passes the drawings to
the checking section. At this point, some drawings may be
required for tracing. The standards section scrutinises the
drawings to ensure that the appropriate standards have been
incorporated. All schedules, equipment lists, and routine
clerical work is normally performed by technical clerks.
Completed work for approval by the chief draughtsman is
returned via the section leader.

Finished drawings are microfilmed and stored. The
print-room staff distribute copies of drawings as required.

Chapter 2

Drawing instruments and materials

Drawing boards

A drawing board is made of well-seasoned wood, generally yellow pine. In most sizes it is fitted with battens at the rear side, to prevent warping, and, in addition, grooves are cut to approximately half the thickness of the board. The working edge of the board is fitted with an insert of ebony, against which the draughtsman slides the stock of the tee-square. The ebony strip provides a harder working edge than the edge of the softer wood could offer.

Drawing boards are available which are manufactured from three-, four-, or thicker multi-ply timber.

Care should be taken with any drawing board to ensure that it is not subjected to extremes of temperature or humidity, or left in direct sunlight for long periods; otherwise warping of the flat surface is bound to take place.

Drawing pins should on no account be used to pin the drawing sheet to the surface of the board. This practice results in obvious damage. Drawing-board clips and also adhesive tape are available, and the use of one or both of these will ensure that the paper is rigidly fixed to the drawing board.

Tee-squares

The tee-square consists of two parts: the stock and the blade. The blade is screwed to the stock, and steady-pins are added. Various woods are used—beech, pear, maple, and mahogany. The side of the stock may also have an ebony edge to slide against the working edge on the drawing board. The working edge of the tee-square should be bevelled. Tee-squares are available with a transparent strip along the working edge; these enable the draughtsman to view part of the drawing beneath the tee-square, which is often advantageous. This feature is also found on drawing boards which are fitted with parallel-slide tee-squares.

Set-squares

Adjustable set-squares are available; squares with the longest side about 250mm are the most serviceable in general use. Lines at any angle can be drawn easily with an adjustable square; alternatively, two set-squares and a protractor will be required. A 60° set-square has angles of 30°, 60°, and 90°; a 45° set-square has angles of 45°, 90°, and 45°. Set-squares are generally made of celluloid and, as mentioned above, large squares are preferrable. Small squares and squares made from wood or metal are not recommended, as the latter hide the drawing.

Drawing instruments

To draw small arcs and radii with a high degree of accuracy, it is necessary to obtain good quality instruments, with a centre adjustment. Instruments are available for both pencil and ink drawing, and fig. 2.1 shows an inexpensive set from the Staedler-Mars range. For larger circles, the compass

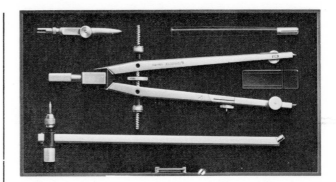

Fig. 2.1

radius may be further increased by the insertion of the horizontal extension bar, which is shown in position in fig. 2.2.

For the very largest of circles, beam compasses are required. These are basically a rigid beam with a sliding head designed to take a pencil point or an ink pen.

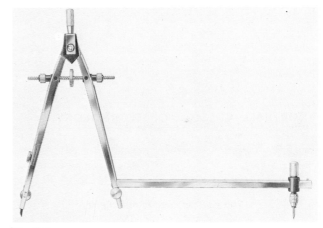

Fig. 2.2

Drawing sheets

Drawings may be produced on paper, linen, or plastics film. These materials are available from a roll or as single sheets cut to size. Drawing sheets are supplied plain or with a printed border and title-block, and examples of prepared sheets appear later in this book. Certain basic information is required on the drawing format, and the details in the title-block used in this book will serve to illustrate some of the more important points. Generally, the information supplied on the prepared sheet will depend on the type of drawing office, the range of products, and the particular design requirements of the individual firm. A typical sheet suitable for a college of technology is shown in fig. 2.3.

To assist in handling, printing, and photographic processing, the sizes of drawing sheets used for normal purposes

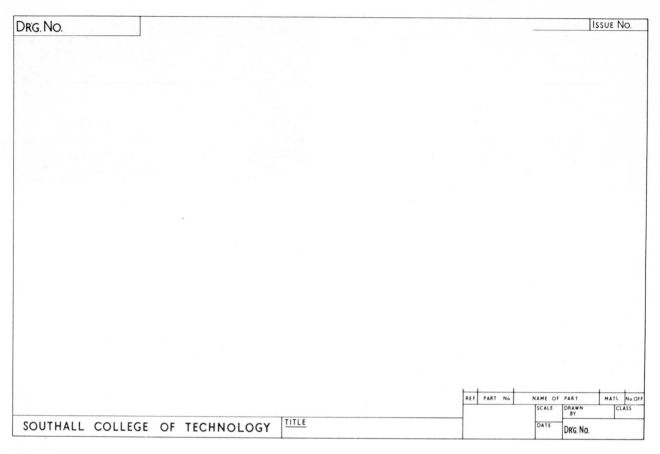

			REF	PART No.	NAME OF PART		MATL	No.OFF
DRG. No.		ISSUE No.			SCALE	DRAWN BY	CLASS	
SOUTHALL COLLEGE OF TECHNOLOGY	TITLE				DATE	DRG. No.		

Fig. 2.3

should be as follows:

Designation	Size in millimetres
A0	841 x 1189
A1	594 x 841
A2	420 x 594
A3	297 x 420
A4	210 x 297

The above sizes include a border which should be a minimum of 15mm and contain, if required, the registration marks for microfilming.

Draughting film

The advantages of draughting film Draughting film offers the following advantages over tracing paper and cloth:

1   superior dimensional stability;
2   more even drawing surface;
3   greater resistance to fraying, breaking, or buckling;
4   no edge-lining necessary;
5   less affected by moisture and climatic changes;
6   perfect transparency—does not yellow or become brittle with age.

Some of the above characteristics did originally create problems when conventional drawing instruments were used, since they were unable to stand up to the greater abrasion created when drawing on film; but the introduction of suitable drawing instruments has enabled the draughtsman to exploit the advantages of film. While film is more expensive than normal tracing paper, this is not an important factor, since well over 90% of the costs of a drawing are due to personnel costs. Considering its advantages, film can be more economical than paper or cloth in many applications.

Types of draughting film The more popular types of film are made of polyester—PVC and polycarbonate films being used to a lesser extent. The matt surface on one or both sides is mostly produced by a chemical coating process, sometimes involving several coatings. In some cases, a matt surface is produced by a mechanical (sand-blasting) or chemical 'roughening up' process of the film surface.

Generally, film of the chemically coated polyester type is preferred. Some suppliers of film buy in basic polyester film and treat the surface to their own processing techniques. Fig. 2.4 shows enlarged cross-sections of different types.

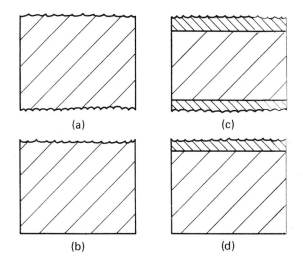

Fig. 2.4   Cross-sections through different types of draughting film. (a) Double-sided film, mechanically or chemically treated (b) Single-sided film, mechanically or chemically treated (c) Double-sided chemically coated film (d) Single-sided chemically coated film

In addition to draughting film with a matt drawing surface, there are glossy types of film chiefly made of acetate. Such film is suitable for special applications connected with surveyors' work or for use in overhead projectors.

The quality of drawings made on draughting film The quality of a drawing made on film depends largely on the following factors:
1 the quality of the film used,
2 the treatment of the film's surface,
3 the use of suitable drawing instruments.
The quality of a drawing on film is also determined by the following properties:
1 firm adhesion of lines drawn to the film surface,
2 clean-cut, non feathered lines,
3 permanence of lines,
4 contrast of lines in relation to the film surface,
5 good erasing properties and ease of correction on erased areas.

The treatment of draughting film before use In order to achieve the best results when draughting on film, it is advisable to clean the surface of the film before starting to draw. For this purpose, the use of cellulose crepe paper, liquid cleaner, or Mars-plastic Eraser no.52652 or similar products has proved to be practical and efficient. A practical test of different liquids is the only way to establish the best type for the film used. Cleaning powder is not ideal for preparing the surface, since it is difficult to remove completely and tends to clog technical pens.

Drawing with pencils and leads on draughting film The use of high-quality graphite drawing pencils and leads can achieve acceptable results on draughting film. High-quality pencils produce black lines which are easy to erase. They are ideal for making guide lines before inking, but there is a danger of lines smudging. Finished drawings, as well as guide lines, can be made with Mars-Dynagraph, since lines produced are smudge proof and, because of their opacity and contrast, reproduce very well.

Draughting on film with drawing ink Ink lines drawn over pencil or lead guide lines should have a broader line width than the guide line. If the guide line is broader than the ink line drawn over it, then the ink deposit will not adhere to the film surface. It should also be noted that drawings for microfilm should on no account be executed partly in graphite or plastic lead and partly in ink, as the difference in contrast is too great. This could create great difficulties in selecting the correct exposure.

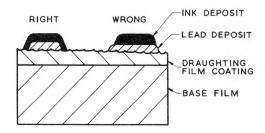

Fig. 2.5   Section through draughting film. For good adhesion, the ink line should be broader than the pencil guide line.

Fig. 2.5 shows a section through film with ink lines. Water-soluble ink is available which will not chip off even if the draughting film is rolled but which can easily be removed by an eraser; etching ink which etches into the draughting-film surface through a chemical reaction between ink and surface is also available. In cases where corrections are absolutely necessary, lines can be removed by applying a liquid solvent. This is not an ideal procedure, since such solvents almost invariably damage the surface of the film.

Chapter 3

Linework and lettering

Lines and linework

Only two thicknesses of line-width are required by current standards for drawing, and careful attention to this detail will result in consistency of presentation. The reader is referred to British Standard BS 308: 1972, 'Engineering drawing practice', where it can clearly be seen that the characteristic features of the linework which are so distinctive are the consistency of the line thicknesses and of their construction where lines are not continuous. Remember when draughting that careful attention to small details at all times will make a big contribution towards a high degree of perfection.

When tracing a drawing in ink, it is a relatively easy matter to regulate line-width, since ruling pens, ink bows, and compasses have a screw adjustment for this purpose. A certain amount of skill is necessary when using a pencil, though, as the circular lead will tend to wear unevenly. A chisel edge can be prepared by rubbing the exposed pencil lead, after sharpening, across a flat sand-paper block, then turning the pencil over and repeating the process in an attempt to produce two parallel faces. However, the best-prepared chisel edge tends to thicken in use just when the draughtsman least expects it to do so, and often in the middle of a long line. All pencils require regular re-sharpening for the draughtsman to produce good linework.

Chuck-type pencils are manufactured with extra strong clutch jaws which grip leads which are available in various thicknesses. These allow drawing, lettering, and writing to be done without the necessity of resharpening; drawing can also be recommenced at any time without line-thickness variations. These thin-lead holders will enable cleaner drawings to be produced, since no sharpening dust exists; and they are also more economical, since no time is lost in sharpening.

Recommended lines are based on two thicknesses—0.7mm and 0.3mm—either as continuous lines or broken to provide dashed and thick- or thin-chain lines.

Care should be taken with dashed lines to ensure that the dashes are of the same length along the line, and regularly spaced. A dash of about 3mm followed by a space of 2mm will produce the line in fig. 3.1.

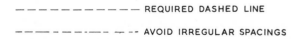

Fig. 3.1

In fig. 3.2 the outline is drawn in a continuous thick line 0.7mm thick. Dashed lines 0.3mm thick are used for hidden detail, and these should begin and end with a dash in contact with the lines from which they start and finish. The preceding mention of 3mm dashes is for guidance only, as obviously some adjustment needs to be made according to

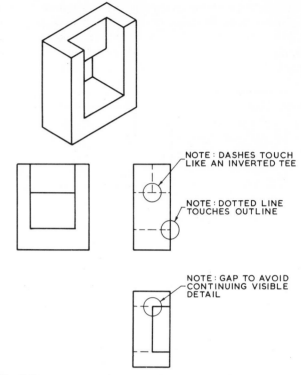

Fig. 3.2

the length of line required for any particular circumstance. Where lines showing hidden detail are an extension of visible outlines, then the dashed line starts with a space. Where dashed lines touch other dashed lines they meet as a letter L, T, or X.

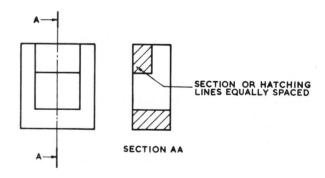

Fig. 3.3

Cutting planes

The chain line in fig. 3.3 indicates a cutting plane. Suppose the component is cut by a hacksaw along this plane; the arrows now indicate which half we are interested in viewing, and the projection angle will decide the position on the paper to present the drawing. Section planes are designated by capital letters.

The metal face assumed to have been cut by the saw is indicated by the section lines, which are continuous. Note that section lines are equally spaced, and should not be closer than 4mm pitch when the drawing is to be microfilmed, since close hatching tends to give a shading effect. When a drawing has many sectioned parts, the experienced draughtsman normally varies the pitch between section lines—increasing them as the parts increase in size, and vice versa.

It will be seen that the section plane has been indicated by a chain line of two different thicknesses: 0.7mm thick at the ends and 0.3mm thick along the central portion. This is a new practice introduced in BS 308: 1972; a thick-chain line is now reserved to indicate a surface which has to meet certain special requirements, for example of flatness and surface finish, as shown in fig. 3.4, which shows part of a shaping machine to summarise the application of various types of line.

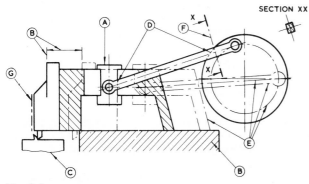

Fig. 3.4
Key: A — drawing outline
     B — dimension lines, projection lines, and cross-hatching
     C — limit of partial view
     D — hidden detail
     E — centres lines and phantom lines indicating an alternative position of the mechanism
     F — cutting plane
     G — indication of a surface requiring a special finish

Centre lines

Centre lines are used to indicate the axes of holes and solids, and they consist of long and short dashes 0.3mm thick. It is impossible to establish dimensions for the long and short dashes, since holes obviously vary considerably in size, and the centre lines should extend only a small distance past the limits of size. Since the function of a centre line is also to indicate the exact position of centres, it should cross another centre line on the solid part of the line and not in a space, as shown in fig. 3.5. The centre

lines should also be arranged in a symmetrical manner in the vicinity of a hole, and they may be extended for the purposes of dimensioning. They should be dimensioned to, but should never themselves be used as dimension lines or to link one view with an adjacent projected view.

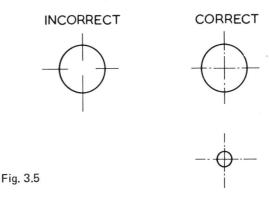

Fig. 3.5

Break lines

It is often necessary to signify a break in a large component where only part is required to be detailed, or to shorten a long part which has uniform shape for all or most of its length. A break line is 0.3mm thick, continuous, but irregular. It should be spaced away from centre lines to avoid confusion. Fig. 3.6 shows examples of both applications.

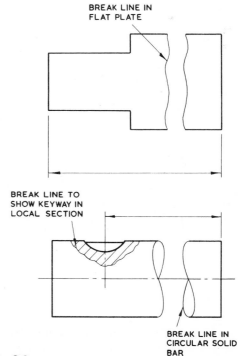

Fig. 3.6

Extension lines and dimension lines

Extension lines and dimension lines, both 0.3mm thick, are shown in fig. 3.7. Note that the extension line does not actually touch the component; it is generally drawn to within about 2mm of the component and the same distance past the dimension line. Consistency in the application of both these details will improve the general appearance of detail drawings where many dimension lines are added.

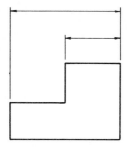

Fig. 3.7

Dimension lines may either be drawn as continuous lines, with the figure printed above the line but not touching it, or broken with the figure printed between the two parts. Where possible, dimension lines should also be positioned outside of the component outline; a drawing where many dimensions are spaced within the outline often has a cramped appearance.

Phantom lines and leader lines

Fig. 3.8 shows a component drawn in two positions, the 'phantom' line being used in this case to indicate an alternative position. A phantom line is a thin chain of long and short dashes proportioned to suit the size of the drawing but 0.3mm thick.

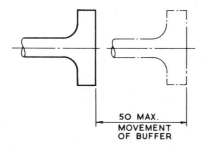

50 MAX.
MOVEMENT
OF BUFFER

Fig. 3.8

A leader line, 0.3mm thick, is used to draw attention to part of a drawing covered by a note or dimension. Fig. 3.9 shows an application.

LEADER LINE — — DATUM SURFACE

Fig. 3.9

Lettering

Fig. 3.10 shows examples of letters of acceptable style which are satisfactory for direct reproduction or for microfilming.

(a) ABCDEFGHIJKLMNOPQRSTUVWXYZ
1234567890/

(b) ABCDEFGHIJKLMNOPQRSTUVWXYZ
1234567890/

(c) ABCDEFGHIJKLMNOPQRSTUVWXYZ
1234567890/

(d) abcdefghijklmnopqrstuvwxyz

(e) 1234567890

(f) 1234567890

Fig. 3.10   (a) Open style (b) Condensed style (c) Sloping style (d) Open style, lower case (e) Open style, freehand (f) Sloping style, freehand

A drawing should have a consistent, uniform style of printing; the use of ink, pencil, stencil, and typescript on the same drawing should be avoided. Printing may be upright or sloping, but not mixed, and should be well spaced, with no unnecessary frills and flourishes. Inconsistent delineation sometimes arises from printing with a chisel-point pencil, and fine lines fail to reproduce well. Care should also be taken with erasures, since, if these are incomplete, it is possible to obtain a ghosted image on the print. Notes do not need to be underlined, but they may be increased in size if considered of extra importance. In addition to hand printing, lettering may be produced by 'Letraset', by machine, or by stencil.

Spacing between lines of print should be at least 50% of the character height for notes on the drawing, but closer spacing of title-blocks is acceptable, since the enlarged print tends to ensure clarity. The use of capital letters is preferred to lower-case letters, since the latter lose some clarity when reduced in size, although lower-case letters are acceptable where they form part of a standard abbreviation or symbol. When used, lower-case letters should be drawn a minimum height of 0.6 times the capital-letter height.

For A1, A2, A3, and A4 sheets the character height should not be less than 2.5mm, and for A0 sheets 3.5mm for dimensions and notes. Drawing numbers for A0, A1, A2, and A3 sheets should be a minimum of 7mm and for A4 sheets 5mm in height. For all characters, the stroke-thickness should be approximately 0.1 times the height of the character, with a clear space of not less than 0.7mm between characters and 2.5mm for capitals. This ensures that, on microfilming, the print does not tend to run into one, but remains clear. All printing should be arranged to be read in the same direction and parallel with the format of the drawing sheet.

The two open styles of lettering shown in fig. 3.10 (e) and (f) are often used on drawings which are to be microfilmed, as increased clarity is obtainable on small reproductions.

Chapter 4

Microfilming

If the draughtsman observes the instructions given in BS 308 regarding line thicknesses, and the drawing is subsequently finished in ink, then the drawing will also be suitable for microfilming. Attention to certain details, however, can render pencil drawings suitable for reproduction by this technique. Microfilming depends on a photographic process, and it is vital that a high degree of contrast exists between finished lines and the paper background. If a drawing is produced with a pencil and only the lettering and dimensioning is inked in, then the pencil line must as far as is possible have the same density as the ink work.

Now the surface of paper is not smooth, but when magnified appears as a random succession of hills and valleys. By passing a pencil across the sheet, a small amount of graphite will be worn away on one side of the hill, and the valley will be mainly left bare. The result is that what appears as a grey line to the naked eye may well consist of 40% black graphite and 60% white paper against a white paper background. Obviously, the softer the pencil, the more graphite will be deposited on the drawing sheet. The maximum density obtainable will quite possibly result from a line which consists of graphite and paper in equal proportions, after using a very soft pencil; hard pencils resist wear, and leave little deposit on the paper.

It should be pointed out that, while the drawing produced with a soft pencil is more apt to smudge, the drawing is often handled far less when microfilmed compared with the drawing which is printed by commercial dyeline machines. Generally, pencils harder than 2H are unsatisfactory, and for the best linework of consistent density a lead one grade softer than that used in the pencil should be used in spring-bows and compasses. Soft pencils also reduce the possibility of paper damage by cutting or scoring of the surface; any deep groove which reflects light leaves a poorly defined line on a microfilm. It will be found beneficial to use a reasonably hard surface as a backing sheet on the drawing board, especially if the drawing paper is relatively soft.

Basic facts on microfilming

The practice of microfilming engineering drawings is increasing in popularity for the following reasons.

1 Protection of costly, and therefore valuable, drawings against damage, fire, theft and, possibly more important, misfiling, and deterioration of drawings due to the combined effects of frequent handling and ageing of vellum or linen originals. If microfilming is properly introduced, the first microfilm of the original drawing is usually mounted on an aperture card; this aperture card is kept in the master file of all drawings. Any number of working drawings is then readily available, as duplicates from the master are easily made. The same holds good for prints obtained from duplicated aperture cards anywhere in the organisation.

2 Space saving. As a rule, aperture cards holding microfilm take up only 5% of the space required by corresponding original drawings.

3 Time saving. Filing and refiling from an aperture-card master file takes up much less time than filing and refiling of original drawings which, if they are large, are sometimes rolled and difficult to handle.

4 Reduced cost of working drawings. This is possibly the most important reason why microfilming is becoming so popular for engineering drawings. As the draughtsman or designer can at any time, and at his work-station, view the drawing or part of the drawing, he will frequently get the required information from the viewer without making a print. There is no doubt that the availability of viewers and viewer/printers has brought about a new, different approach to microfilming. It has been in use for archival purposes for some time, but the viewer/printer has now made microfilming also an ideal means of reproduction or retrieval of engineering information. Another reason why printing costs are reduced in a microfilming system is that the draughtsman who knows that a master file is readily available will not tend to collect too many drawings at his work-station.

5 Inexpensive distribution. This is of importance with larger firms operating engineering subsidiaries in different places or countries; it is obviously cheaper and safer to mail aperture cards rather than printed sheets.

6 Microfilming facilitates changes and additions to drawings. In many drawing offices, changes are a major operation, in which draughtsmen need to go back to the original drawings. On the other hand, original drawings should not be changed, because each stage of the design of a product might be needed for service requirements in the future. Under the old system, expensive master originals had to be retained for each little change, whereas now the same effect is achieved, at a much lower cost, by microfilming each stage.

7 Drawing to scale. Microfilming can save the reproduction of a drawing to another scale.

For the reasons stated, many large organisations have gone over or are going over to microfilming, and, as a result, they often ask their sub-contractors to produce drawings to microfilm standards. For this reason, microfilming is not only of interest to larger firms but can be critically important to smaller firms. Another point to consider is that drawings generally have a fairly long life. It can be safely assumed that microfilming, be it on the premises of a firm, or outside, on a time-sharing basis, will be introduced into most firms during the next few years. It is logical, therefore, to draw to microfilm standards now, even if the originals are not to be microfilmed immediately.

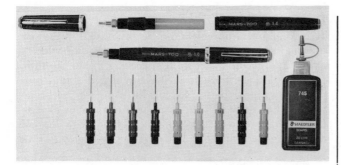

Fig. 4.1   Technical pens for lines of constant width

Draughting techniques for microfilming

Both paper and draughting film can be used for originals. Translucency is not as essential as it is with other reproduction processes; as stated before, it is the contrast between the line and the surroundings which counts in microfilming, hence even relatively cheap paper can be used. However, most accurate results, especially when higher reduction-ratios are considered, are obtained from using draughting film which is dimensionally stable under all conditions. Since a drawing may take weeks to complete, and as the accuracy of it is influenced by expansion and contraction of the drawing surface, the extra expense of draughting film is sometimes considered worthwhile—the more so as tolerances in the accuracy of the drawing are compounded by the reduction in microfilming and subsequent re-enlargement to greater than original size.

There are no hard and fast rules regarding the use of ink or pencil on original drawings to be microfilmed, as the decision depends entirely on the type of drawing, reduction-ratios, film resolution, and film density required. The best results, from the point of view of quality, are obtained from ink drawings. They offer perfect contrast and, provided technical pens are used, line-widths can be conveniently controlled. Fig. 4.1 shows typical technical pens for microfilm drawing. For consistency of line thickness, technical pens each producing only a single line conforming to the standard dimension have advantages over an adjustable ruling pen. Originals to be microfilmed should preferably not be partly in pencil and partly in ink, as the difference in the contrast may be too great, and this poses great difficulties in selecting the correct exposure.

Fig. 4.2 shows the Planvu microfilm reader, manufactured by Lepton Engineering, Chalfont St Peter, Bucks, which accepts 35mm roll film, aperture cards, or strip; and is so

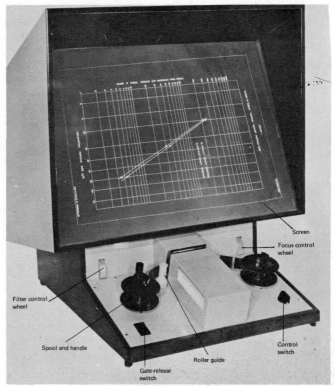

Fig. 4.2

designed that card can be fed from either side to a definite stop which accurately locates the aperture in the centre of the optics. The reader has a high degree of enlargement and, with a magnification of 15 times, the necessity for image orientation is eliminated and film of AO size drawings can be clearly read. The optical system gives full illumination over a large screen measuring 470 x 665mm. A high intensity of illumination is provided by a 150 watt quartz-iodine lamp. This light intensity can be controlled from a three-position switch and this facility allows negatives with poor definition to be read with ease, and also prolongs the life of the lamp. Overheating of the lamp and film is prevented by a quiet-running cooling fan. A green filter is also fitted in order to avoid eye strain when reading film of books or similar documents for prolonged periods.

This equipment can be used effectively in normal lighting but has the additional advantage that in darkened surroundings an image can also be projected onto a suitable screen.

Chapter 5

Scales

Small objects are sometimes drawn larger than actual size, while large components and assemblies, of necessity, are drawn to a reduced size. A drawing should always state the scale used, the scale of a full-size drawing being indicated as 1:1. Drawings themselves should not be scaled when in use for manufacturing purposes, and warnings against the practice are often quoted on standard drawing sheets, e.g. 'DO NOT SCALE' and 'IF IN DOUBT, ASK'. A drawing must be adequately dimensioned, or referenced sufficiently so that all sizes required are obtainable.

The recommended multipliers for scale drawings are 2, 5, and 10.

1:1 denotes a drawing drawn full-size.
2:1 denotes a drawing drawn twice full-size.
5:1 denotes a drawing drawn five times full size.

Other common scales are 10:1, 20:1, 50:1, 100:1, 200:1, 500:1, and 1000:1.

It should be pointed out that a scale drawing can be deceiving; a component drawn twice full-size will cover four times the area of drawing paper as the same component drawn full-size, and its actual size may be difficult to visualise. To assist in appreciation, it is a common practice to add a pictorial view drawn full-size, provided that the drawing itself is intended to be reproduced to the same scale and not reproduced and reduced by microfilming.

The recommended divisors for scale drawings are also 2, 5, and 10.

1:1 denotes a drawing drawn full-size
1:2 denotes a drawing drawn half full-size.
1:5 denotes a drawing drawn a fifth full-size.

Other common scales used are 1:10, 1:20, 1:50, 1:100, 1:200, 1:500, and 1:1000.

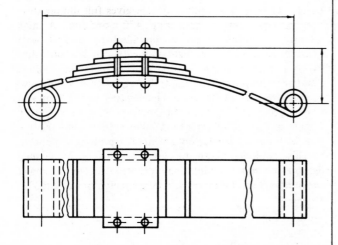

Fig. 5.1   Example of dual scales—co-ordinate dimensions 1:5, superimposed detail 1:2

Dual scales

It is sometimes convenient to use two scales on the same drawing; the example in fig. 5.1 shows a spring with the length and height drawn to a scale of 1:5 and the spring assembly 1:2. Drawing in dual scales is also a convenient practice when preparing foundation plans for mechanical plant and equipment; the centres between groups of fixing holes can be drawn to a convenient scale, with the hole details for the foundation bolts drawn to an enlarged scale.

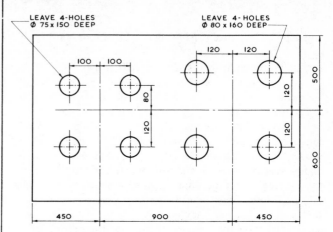

Fig. 5.2   Plan of concrete plinth, showing use of dual scales. Scale for hole details—1:5. Scale for plinth details—1:10.

Division of lines

Fig. 5.3 shows the method of dividing a given line AB, 89mm long, into a number of parts (say 7).

Draw line AC, and measure 7 equal divisions. Draw line B7, and with the tee-square and set-square draw lines parallel to line B7 through points 1 to 6, to give the required divisions on AB.

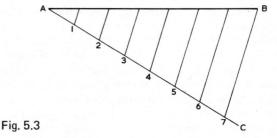

Fig. 5.3

Fig. 5.4 shows an alternative method.
1 Draw vertical lines from A and B.
2 Place the scale rule across the vertical lines so that seven equal divisions are obtained and marked.
3 Draw vertical lines up from points 2 to 7 to intersect AB.

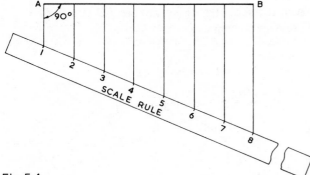

Fig. 5.4

Diagonal scales

Fig. 5.5 shows the method of drawing a diagonal scale of 40mm to 1m which can be read by 10mm to 4m. Diagonal scales are so called since diagonals are drawn in the rectangular part at the left-hand end of the scale. The diagonals produce a series of similar triangles.

1 Draw a line 160mm long.
2 Divide the line into four equal parts.
3 Draw 10 vertical divisions as shown and to any reasonable scale (say 5mm) and add diagonals.

An example of reading the scale is given.

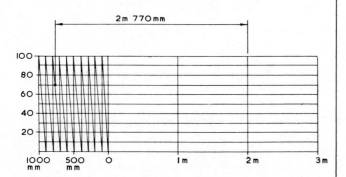

Fig. 5.5   Diagonal scale where 40mm represents 1m

Plain scales

The method of drawing a plain scale is shown in fig. 5.6. The example is for a plain scale of 30mm to 500mm to read by 125mm to 2.5m.

1 Draw a line 150mm long and divide it into 5 equal parts.
2. Divide the first 30mm length into four equal parts, and note the zero position on the solution.

An example of a typical reading is given.

This method of calibration is in common use in industry, and scales can be obtained suitable for a variety of scale ratios.

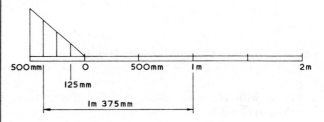

Fig. 5.6   Plain scale where 30mm represents 500mm

Chapter 6

Practical geometry

To bisect a given angle AOB (fig. 6.1)

1　With centre O, draw an arc to cut OA at C and OB at D.
2　With centres C and D, draw equal radii to intersect at E.
3　Line OE bisects angle AOB.

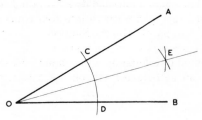

Fig. 6.1

To bisect a given straight line AB (fig. 6.2)

1　With centre A and radius greater than half AB, describe an arc.
2　Repeat with the same radius from B, the arcs intersecting at C and D.
3　Join C to D and this line will be perpendicular to and bisect AB.

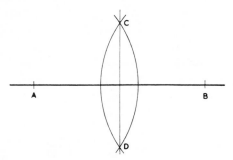

Fig. 6.2

To bisect a given arc AB (fig. 6.3)

1　With centre A and radius greater than half AB, describe an arc.
2　Repeat with the same radius from B, the arcs intersecting at C and D.
3　Join C to D to bisect the arc AB.

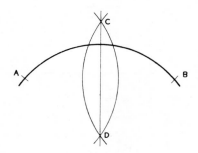

Fig. 6.3

To find the centre of a given arc AB (fig. 6.4)

1　Draw two chords, AC and BD.
2　Bisect AC and BD as shown in fig. 6.2; the bisectors will intersect at E.
3　The centre of the arc is point E.

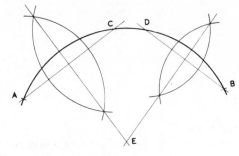

Fig. 6.4

To inscribe a circle in a given triangle ABC (fig. 6.5)

1　Bisect any two of the angles as shown in fig. 6.1, so that the bisectors intersect at D.
2　The centre of the inscribed circle is point D.

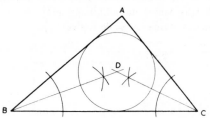

Fig. 6.5

To circumscribe a circle around triangle ABC (fig. 6.6)

1　Bisect any two of the sides of the triangle as shown in fig. 6.2, so that the bisectors intersect at D.
2　The centre of the circumscribing circle is point D.

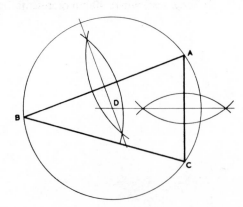

Fig. 6.6

To draw a hexagon, given the distance across the corners

Method A (fig. 6.7 (a))

1 Draw vertical and horizontal centre lines and a circle with a diameter equal to the given distance.
2 Step off the radius around the circle to give six equally spaced points, and join the points to give the required hexagon.

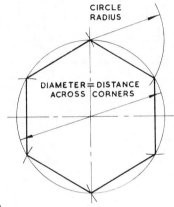

Fig. 6.7 (a)

Method B (fig. 6.7 (b))

1 Draw vertical and horizontal centre lines and a circle with a diameter equal to the given distance.
2 With a 60° set-square, draw points on the circumference 60° apart.
3 Connect these six points by straight lines to give the required hexagon.

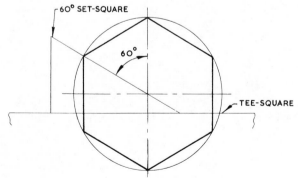

Fig. 6.7 (b)

To draw a hexagon, given the distance across the flats (fig. 6.8)

1 Draw vertical and horizontal centre lines and a circle with a diameter equal to the given distance.
2 Use a 60° set-square and tee-square as shown, to give the six sides.

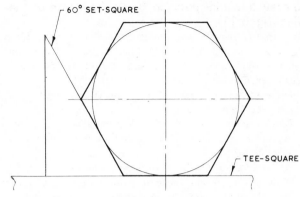

Fig. 6.8

To draw a regular octagon, given the distance across corners (fig. 6.9)

Repeat the instructions in fig. 6.7 (b), but use a 45° set-square, then connect the eight points to give the required octagon.

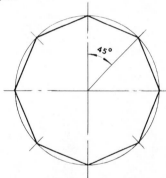

Fig. 6.9

To draw a regular octagon, given the distance across the flats (fig. 6.10)

Repeat the instructions in Fig. 6.8, but use a 45° set-square to give the required octagon.

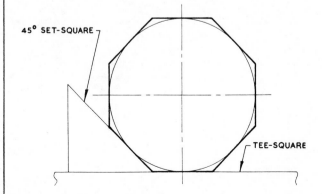

Fig. 6.10

To draw a regular polygon, given the length of the sides (fig. 6.11)

Note that a regular polygon is defined as a plane figure which is bounded by straight lines of equal length and which contains angles of equal size. Assume the number of sides is seven in this example.

1 Draw the given length of one side AB, and with radius AB describe a semi-circle.
2 Divide the semi-circle into seven equal angles, using a protractor, and through the second division from the left join line A2.
3 Draw radial lines from A through points 3, 4, 5, and 6.
4 With radius AB and centre on point 2, describe an arc to meet the extension of line A3, shown here as point F.
5 Repeat with radius AB and centre F to meet the extension of line A4 at E.
6 Connect the points as shown, to complete the required polygon.

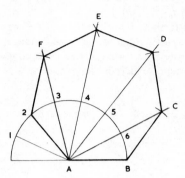

Fig. 6.11

Chapter 7

Tangency

If a disc stands on its edge on a flat surface it will touch the surface at one point. This point is known as the point of tangency, as shown in fig. 7.1, and the straight line which represents the flat plane is known as a tangent. A line drawn from the point of tangency to the centre of the disc is called a normal, and the tangent makes an angle of 90° with the normal.

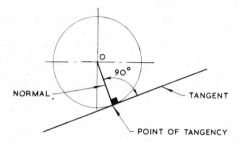

Fig. 7.1

The following constructions show the methods of drawing tangents in various circumstances.

1 To draw a tangent to a point A on the circumference of a circle, centre O (fig. 7.2)

Join OA and extend the line for a short distance. Erect a perpendicular at point A by the method shown.

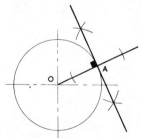

Fig. 7.2

2 To draw a tangent to a circle from any given point A outside the circle (fig. 7.3)

Join A to the centre of the circle O. Bisect line AO so that point B is the mid-point of AO. With centre B, draw a semi-circle to intersect the given circle at point C. Line AC is the required tangent.

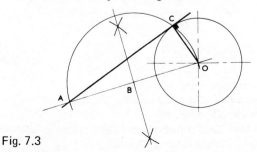

Fig. 7.3

3 To draw an external tangent to two circles (fig. 7.4)

Join the centres of the circles by line AB, bisect AB, and draw a semi-circle. Position point E so that DE is equal to the radius of the smaller circle. Draw radius AE to cut the semi-circle at point G. Draw line AGH so that H lies on the circumference of the larger circle. Note that angle AGB lies in a semi-circle and will be 90° Draw line HJ parallel to BG. Line HJ will be tangential to the two circles and lines BJ and AGH are the normals.

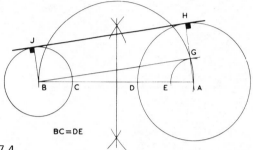

BC=DE

Fig. 7.4

4 To draw an internal tangent to two circles (fig. 7.5)

Join the centres of the circles by line AB, bisect AB and draw a semi-circle. Position point E so that DE is equal to the radius of the smaller circle BC. Draw radius AE to cut the semi-circle in H. Join AH; this line crosses the larger circle circumference at J. Draw line BH. From J draw a line parallel to BH to touch the smaller circle at K. Line JK is the required tangent. Note that angle AHB lies in a semi-circle and will therefore be 90°. AJ and BK are normals.

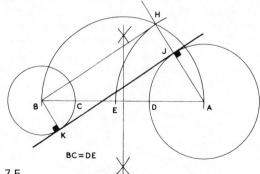

BC=DE

Fig. 7.5

5 To draw internal and external tangents to two circles of equal diameter (fig. 7.6)

Join the centres of both circles by line AB. Erect perpendiculars at points A and B to touch the circumferences of the circles at points C and D. Line CD will be the external tangent. Bisect line AB to give point

E, then bisect BE to give point G. With radius BG, describe a semi-circle to cut the circumference of one of the given circles at H. Join HE and extend it to touch the circumference of the other circle at J. Line HEJ is the required tangent. Note that again the angle in the semi-circle, BHE, will be 90°, and hence BH and AJ are normals.

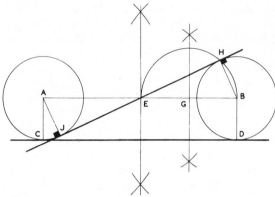

Fig. 7.6

6 To draw a curve of given radius to touch two circles when the circles are outside the radius (fig. 7.7)

Assume that the radii of the given circles are 20 and 25mm, spaced 85mm apart, and that the radius to touch them is 40mm.

With centre A, describe an arc equal to 20 + 40 = 60mm.

With centre B, describe an arc equal to 25 + 40 = 65mm.

The above arcs intersect at point C. With a radius of 40mm, describe an arc from point C as shown, and note that the points of tangency between the arcs lie along the lines joining the centres AC and BC. It is particularly important to note the position of the points of tangency before lining in engineering drawings, so that the exact length of an arc can be established.

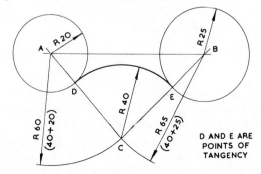

Fig. 7.7

7 To draw a curve of given radius to touch two circles when the circles are inside the radius (fig. 7.8)

Assume that the radii of the given circles are 22 and 26mm, spaced 86mm apart, and that the radius to touch them is 100mm.

With centre A, describe an arc equal to 100 − 22 = 78mm.

With centre B, describe an arc equal to 100 − 26 = 74mm.

The above arcs intersect at point C. With a radius of 100mm, describe an arc from point C, and note that in this case the points of tangency lie along line CA extended to D and along line CB extended to E.

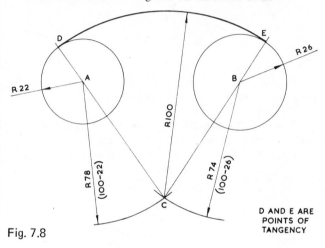

Fig. 7.8

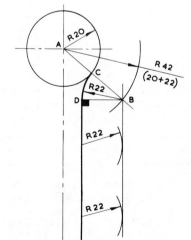

Fig. 7.9

8 To draw a radius to join a straight line and a given circle (fig. 7.9)

Assume that the radius of the given circle is 20mm and that the joining radius is 22mm.

With centre A, describe an arc equal to 20 + 22 = 42mm.

Draw a line parallel to the given straight line and at a perpendicular distance of 22mm from it, to intersect the arc at point B.

With centre B, describe the required radius of 22mm, and note that one point of tangency lies on the line AB at C; the other lies at point D such that BD is at $90°$ to the straight line.

9 To draw a radius which is tangential to given straight lines (fig. 7.10)

Assume that a radius of 25mm is required to touch the lines shown in the figures. Draw lines parallel to the given straight lines and at a perpendicular distance of 25mm from them to intersect at points A. As above, note that the points of tangency are obtained by drawing perpendiculars through the point A to the straight lines in each case.

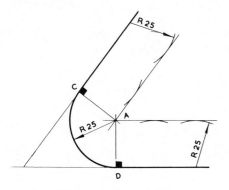

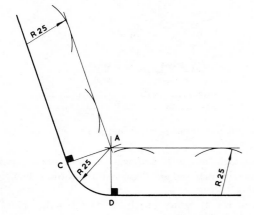

Fig. 7.10

Chapter 8

Loci

If a point, line, or surface moves according to a mathematically defined condition, then a curve known as a *locus* is formed. The following examples of curves and their constructions are widely used and applied in all types of engineering.

Methods of drawing an ellipse

1 Two-circle method Construct two concentric circles equal in diameter to the major and minor axes of the required ellipse. Let these diameters be AB and CD in fig. 8.1.

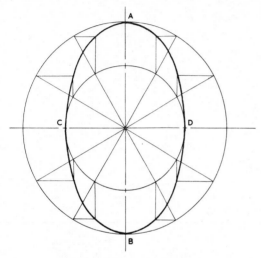

Fig. 8.1   Two-circle construction for an ellipse

Divide the circles into any number of parts; the parts do not necessarily have to be equal. The radial lines now cross the inner and outer circles.

Where the radial lines cross the outer circle, draw short lines parallel to the minor axis CD. Where the radial lines cross the inner circle, draw lines parallel to AB to intersect with those drawn from the outer circle. The points of intersection lie on the ellipse. Draw a smooth connecting curve.

2 Trammel method Draw major and minor axes at right angles, as shown in fig. 8.2.

Take a strip of paper for a trammel and mark on it half the major and minor axes, both measured from the same end. Let the points on the trammel be E, F, and G.

Position the trammel on the drawing so that point F always lies on the major axis AB and point G always lies on the minor axis CD. Mark the point E with each position of the trammel, and connect these points to give the required ellipse.

Note that this method relies on the difference between half the lengths of the major and minor axes, and where these

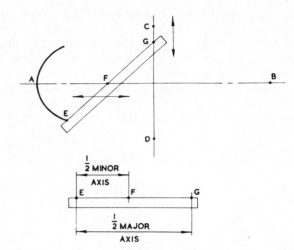

Fig. 8.2   Trammel method for ellipse construction

axes are nearly the same in length, it is difficult to position the trammel with a high degree of accuracy. The following alternative method can be used.

Draw major and minor axes as before, but extend them in each direction as shown in fig. 8.3.

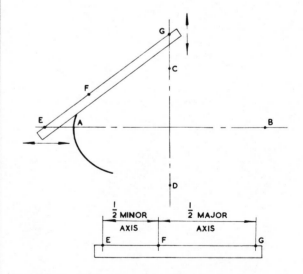

Fig. 8.3   Alternative trammel method

Take a strip of paper and mark half of the major and minor axes in line, and let these points on the trammel be E, F, and G.

Position the trammel on the drawing so that point G always moves along the line containing CD; also, position point E along the line containing AB. For each position of the trammel, mark point F and join these points with a smooth curve to give the required ellipse.

3 To draw an ellipse using the two foci Draw major and minor axes intersecting at point O, as shown in fig. 8.4. Let these axes be AB and CD. With a radius equal to half the major axis AB, draw an arc from centre C to intersect AB at points F_1 and F_2. These two points are the foci. For any ellipse, the sum of the distances PF_1 and PF_2 is a constant, where P is any point on the ellipse. The sum of the distances is equal to the length of the major axis.

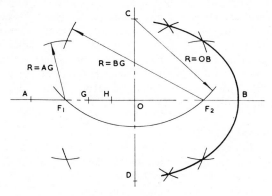

Fig. 8.4   Ellipse by foci method

Divide distance OF_1 into equal parts. Three are shown here, and the points are marked G and H.

With centre F_1 and radius AG, describe an arc above and beneath line AB.

With centre F_2 and radius BG, describe an arc to intersect the above arcs.

Repeat these two steps by firstly taking radius AG from point F_2 and radius BG from F_1.

The above procedure should now be repeated using radii AH and BH. Draw a smooth curve through these points to give the ellipse.

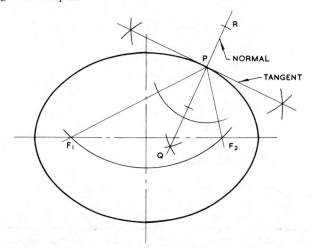

Fig. 8.5

It is often necessary to draw a tangent to a point on an ellipse. In fig. 8.5, P is any point on the ellipse, and F_1 and F_2 are the two foci. Bisect angle F_1PF_2 with line QPR. Erect a perpendicular to line QPR at point P, and this will be a tangent to the ellipse at point P.

The methods of drawing ellipses illustrated above are all accurate. Approximate ellipses can be constructed as follows.

Approximate method 1 Draw a rectangle with sides equal in length to the major and minor axes of the required ellipse, as shown in fig. 8.6.

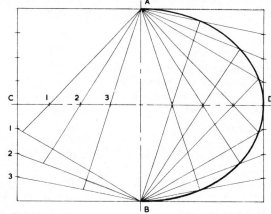

Fig. 8.6

Divide the major axis into an equal number of parts; eight parts are shown here. Divide the side of the rectangle into the same equal number of parts. Draw a line from A through point 1, and let this line intersect the line joining B to point 1 at the side of the rectangle as shown. Repeat for all other points in the same manner, and the resulting points of intersection will lie on the ellipse.

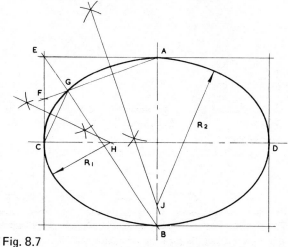

Fig. 8.7

Approximate method 2 Draw a rectangle with sides equal to the lengths of the major and minor axes, as shown in fig. 8.7.

Bisect EC to give point F. Join AF and BE to intersect at point G. Join CG. Draw the perpendicular bisectors of lines CG and GA, and these will intersect the centre lines at points H and J.

Using radii CH and JA, the ellipse can be constructed by using four arcs of circles.

The involute

The involute is defined as the path of a point on a straight line which rolls without slip along the circumference of a cylinder. The involute curve will be required in a later chapter for the construction of gear teeth.

Involute construction

1 Draw the given base circle and divide it into, say, 12 equal divisions as shown in fig. 8.8. Generally only the first part of the involute is required, so the given diagram shows a method using half of the length of the circumference.
2 Draw tangents at points 1, 2, 3, 4, 5, and 6.
3 From point 6, mark off a length equal to half the length of the circumference.
4 Divide line 6G into six equal parts
5 From point 1, mark point B such that 1B is equal to one part of line 6G.
6 From point 2, mark point C such that 2C is equal to two parts of line 6G.

Repeat the above procedure from points 3, 4, and 5, increasing the lengths along the tangents as before by one part of line 6G.
7 Join points A to G, to give the required involute.

Alternative method

1 As above, draw the given base circle, divide into, say, 12 equal divisions, and draw the tangents from points 1 to 6.
2 From point 1 and with radius equal to the chordal length from point 1 to point A, draw an arc terminating at the tangent from point 1 at point B.
3 Repeat the above procedure from point 2 with radius 2B terminating at point C.
4 Repeat the above instructions to obtain points D, E, F, and G, and join points A to G to give the required involute.

The alternative method given is an approximate method, but is reasonably accurate provided that the arc length is short; the difference in length between the arc and the chord introduces only a minimal error.

Archimedean spiral

The Archimedean spiral is the locus of a point which moves around a centre at uniform angular velocity and at the same time moves away from the centre at uniform linear velocity. The construction is shown in fig. 8.9.

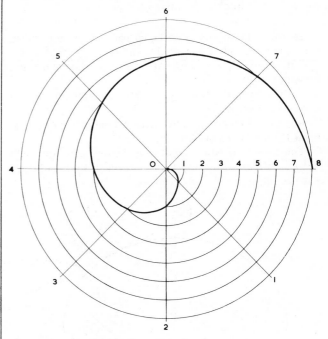

Fig. 8.9   Archimedean spiral

1 Given the diameter, divide the circle into an even number of divisions and number them.
2 Divide the radius into the same number of equal parts.
3 Draw radii as shown to intersect radial lines with corresponding numbers, and connect points of intersection to give the required spiral.

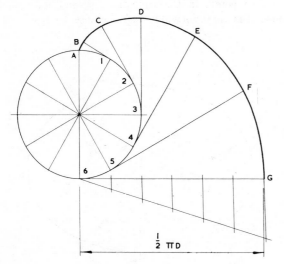

Fig. 8.8   Involute construction

Note that the spiral need not start at the centre; it can start at any point along a radius, but the divisions must be equal.

Self-centring lathe chucks utilise Archimedean spirals.

Right-hand cylindrical helix

The helix is a curve generated on the surface of the cylinder by a point which revolves uniformly around the cylinder and at the same time either up or down its surface. The method of construction is shown in fig. 8.10.

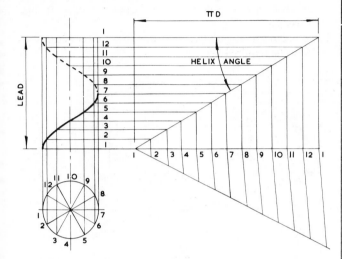

Fig. 8.10   Right-hand cylindrical helix

1. Draw the front elevation and plan views of the cylinder, and divide the plan view into a convenient number of parts (say 12) and number them as shown.
2. Project the points from the circumference of the base up to the front elevation.
3. Divide the lead into the same number of parts as the base, and number them as shown.
4. Draw lines of intersection from the lead to correspond with the projected lines from the base.
5. Join the points of intersection, to give the required cylindrical helix.
6. If a development of the cylinder is drawn, the helix will be projected as a straight line. The angle between the helix and a line drawn parallel with the base is known as the helix angle.

Note. If the numbering in the plan view is taken in the clockwise direction from point 1, then the projection in the front elevation will give a left-hand helix.

The construction for a helix is shown applied to a right-hand helical spring in fig. 8.11. The spring is of square cross-section, and the four helices are drawn from the two outside corners and the two corners at the inside diameter. The pitch of the spring is divided into 12 equal parts, to correspond with the 12 equal divisions of the circle in the end elevation, although only half of the circle need be drawn. Points are plotted as previously shown.

A single-start square thread is illustrated in fig. 8.12. The construction is similar to the previous problem, except that the centre is solid metal. Four helices are plotted, spaced as shown, since the thread-width is half the pitch.

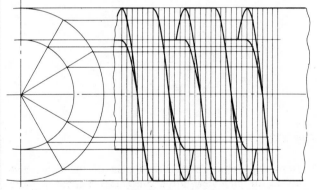

Fig. 8.12   Single-start square thread

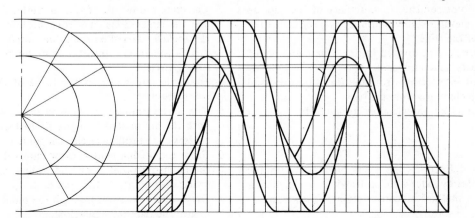

Fig. 8.11   Square-section right-hand helical spring

Right-hand conical helix

The conical helix is a curve generated on the surface of the cone by a point which revolves uniformly around the cone and at the same time either up or down its surface. The method of construction is shown in fig. 8.13.

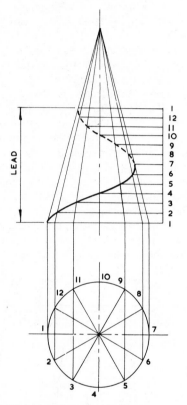

Fig. 8.13   Right-hand conical helix

1 Draw the front elevation and plan of the cone, and divide the plan view into a convenient number of parts (say 12) and number them as shown.
2 Project the points on the circumference of the base up to the front elevation, and continue the projected lines to the apex of the cone.
3 The lead must now be divided into the same number of parts as the base, and numbered.
4 Draw lines of intersection from the lead to correspond with the projected lines from the base.
5 Join the points of intersection, to give the required conical helix.

The cycloid

The cycloid is defined as the locus of a point on the circumference of a cylinder which rolls without slip along a flat surface. The method of construction is shown in fig. 8.14.

1 Draw the given circle, and divide into a convenient number of parts; eight divisions are shown in fig. 8.14.
2 Divide line AA_1 into eight equal lengths. Line AA_1 is equal to the length of the circumference.
3 Draw vertical lines from points 2 to 8 to intersect with the horizontal line from centre O at points O_2, O_3, etc.
4 With radius OA and centre O_2, describe an arc to intersect with the horizontal line projected from B.
5 Repeat with radius OA from centre O_3 to intersect with the horizontal line projected from point C. Repeat this procedure.
6 Commencing at point A, join the above intersections to form the required cycloid.

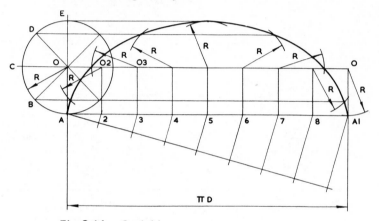

Fig. 8.14   Cycloid

The epicycloid

An epicycloid is defined as the locus of a point on the circumference of a circle which rolls without slip around the outside of another circle. The method of construction is shown in fig. 8.15.

1 Draw the curved surface and the rolling circle, and divide the circle into a convenient number of parts (say 6) and number them as shown.
2 Calculate the length of the circumference of the smaller and the larger circle, and from this information calculate the angle θ covered by the rolling circle.
3 Divide the angle θ into the same number of parts as the rolling circle.
4 Draw the arc which is the locus of the centre of the rolling circle.
5 The lines forming the angles in step 3 will now intersect with the arc in step 4 to give six further positions of the centres of the rolling circle as it rotates.
6 From the second centre, draw radius R to intersect with the arc from point 2 on the rolling circle. Repeat this process for points 3, 4, 5, and 6.
7 Draw a smooth curve through the points of intersection, to give the required epicycloid.

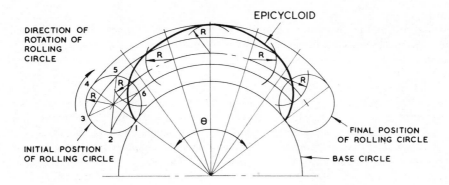

Fig. 8.15   Epicycloid

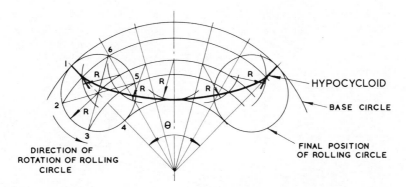

Fig. 8.16   Hypocycloid

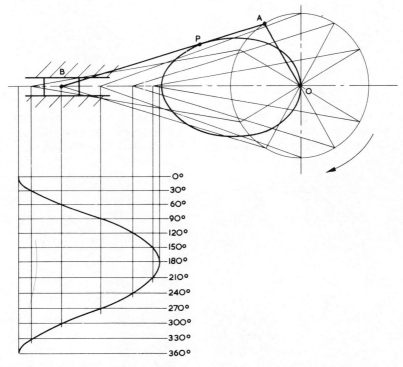

Fig. 8.17

The hypocycloid

A hypocycloid is defined as the locus of a point on the circumference of a circle which rolls without slip around the inside of another circle.

The construction for the hypocycloid (fig. 8.16) is very similar to that for the epicycloid, but note that the rolling circle rotates in the opposite direction for this construction.

It is often necessary to study the paths taken by parts of oscillating, reciprocating, or rotating mechanisms; from a knowledge of displacement and time, information regarding velocity and acceleration can be obtained. It may also be required to study the extreme movements of linkages, so that safety guards can be designed to protect machine operators.

Fig. 8.17 shows a crank OA, a connecting rod AB, and a piston B which slides along the horizontal axis BO. P is any point along the connecting rod. To plot the locus of point P, a circle of radius OA has been divided into twelve equal parts. From each position of the crank, the connecting rod

is drawn, distance AP measured, and the path taken for one revolution lined in as indicated.

The drawing also shows the piston-displacement diagram. A convenient vertical scale is drawn for the crank angle, and in this case clockwise rotation was assumed to start from the 9 o'clock position. From each position of the piston, a vertical line is drawn down to the corresponding crank-angle line, and the points of intersection are joined to give the piston-displacement diagram.

The locus of the point P can also be plotted by the trammel method indicated in fig. 8.18. Point P_1 can be marked for any position where B_1 lies on the horizontal line, provided A_1 also lies on the circumference of the circle radius OA. This method of solving some loci problems has the advantage that an infinite number of points can easily be obtained, and these are especially useful where a change in direction in the loci curve takes place.

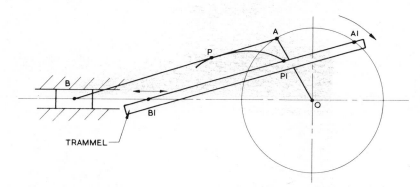

Fig. 8.18

Fig. 8.19 shows a crank OA rotating anticlockwise about centre O. A rod BC is connected to the crank at point A, and this rod slides freely through a block which is allowed to pivot at point S. The loci of points B and C are indicated after reproducing the mechanism in 12 different positions. A trammel method could also be used here if required.

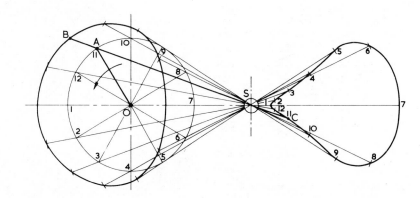

Fig. 8.19

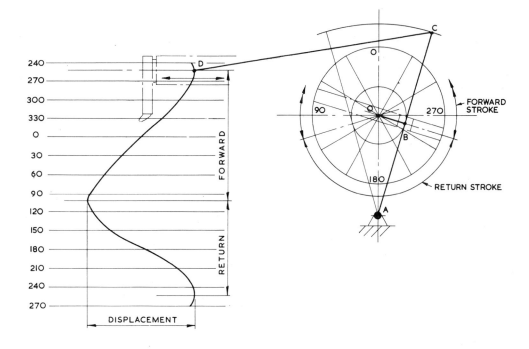

Fig. 8.20

Part of a shaping-machine mechanism is given in fig. 8.20. Crank OB rotates about centre O. A is a fixed pivot point, and CA slides through the pivoting block at B. Point C moves in a circular arc of radius AC, and is connected by link CD, where point D slides horizontally. In the position shown, angle OBA is 90°, and if OB now rotates anti-clockwise at constant speed it will be seen that the forward motion of point D takes more time than the return motion. A displacement diagram for point D has been constructed as previously described.

In fig. 8.21 the radius OB has been increased, with the effect of increasing the stroke of point D. Note also that the return stroke in this condition is quicker than before.

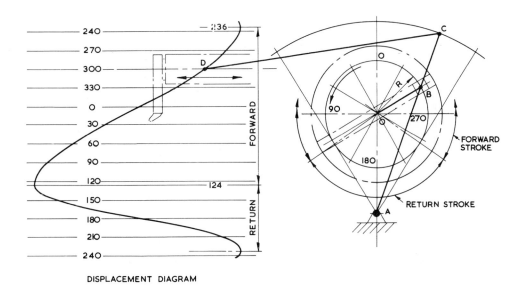

DISPLACEMENT DIAGRAM

Fig. 8.21

The outlines of two gears are shown in fig. 8.22, where the pitch circle of the larger gear is twice the pitch circle of the smaller gear. As a result, the smaller gear rotates twice while the larger gear rotates once. The mechanism has been drawn in twelve positions to plot the path of the pivot point C, where links BC and CA are connected. A trammel method cannot be applied successfully in this type of problem.

Fig. 8.23 gives an example of Watt's straight-line motion. Two levers AX and BY are connected by a link AB, and the plotted curve is the locus of the mid-point P. The levers in this instance oscillate in circular arcs. This mechanism was used in engines designed by James Watt, the famous engineer.

A toggle action is illustrated in fig. 8.24, where a crank rotates anticlockwise. Links AC, CD, and CE are pivoted at C. D is a fixed pivot point, and E slides along the horizontal axis. The displacement diagram has been plotted as previously described, but note that, as the mechanism at E slides to the right, it is virtually stationary between points 9, 10, and 11.

The locus of any point B is also shown.

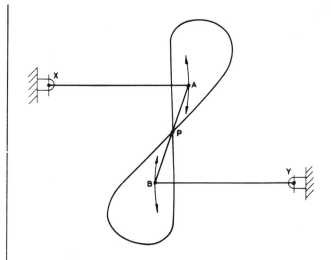

Fig. 8.23   Watt's straight-line motion

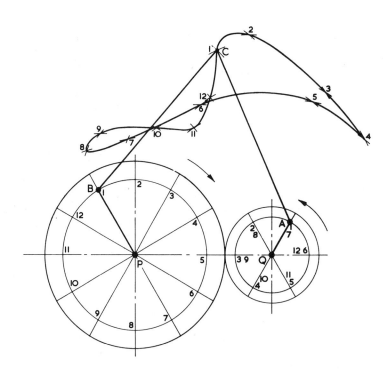

Fig. 8.22

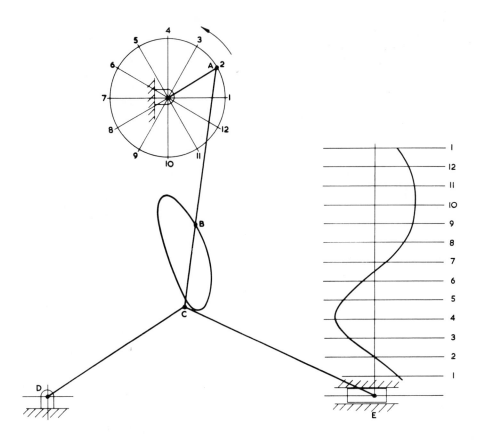

Fig. 8.24

Chapter 9

First- and third-angle projection

First-angle (European) projection

To present the dimensions of, for example, a tapered roller on a working drawing, two alternative methods are internationally used. In fig. 9.1 below, an arbitrary view is taken in line with arrow B. Note that the roller axis lies parallel to the plane of the paper and that the circles at each end of the roller are at 90° to the surface of the paper. If the roller is viewed along its axis in line with arrow A, then it is possible to draw the view obtained in two positions on the paper, namely to the right or to the left of the view previously obtained from arrow B, and in projection with this view.

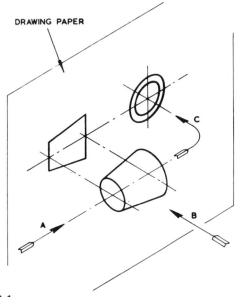

Fig. 9.1

Fig. 9.1 shows the view from arrow A positioned on the remote right-hand side. This method of projection is known as *first-angle* or *European projection*, and is widely practiced in the United Kingdom, on the Continent, and in the USSR.

Third-angle (American) projection

Fig. 9.2 shows the alternative position of the view from arrow A, on the left-hand side, and arrows C and D indicate the way the view from arrow A needs to be rotated in order to position itself on the adjacent side. This method of projection is known as *third-angle* or *American projection*, and is also practiced in the United Kingdom.

Figs 9.1 and 9.2 both show the roller viewed from the left-hand side. Obviously the roller could have been viewed from above, from beneath, or from the right-hand side, giving five different views in either first- or third-angle projection.

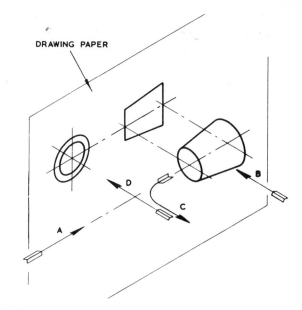

Fig. 9.2

BS 308 accepts that both first- and third-angle projection are internationally accepted methods of presentation, but it stipulates that a drawing should include a symbol indicating which system has been used, to avoid confusion (figs 9.3 and 9.4).

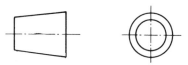

Fig. 9.3   First-angle projection drawing symbol

Fig. 9.4   Third-angle projection drawing symbol

Drawing in first-angle projection

Fig. 9.5 shows a pictorial view of a small component which is to be used to illustrate the principle of first-angle projection.

Unless an article can be oriented to give a definite front view, the draughtsman has no option but to make an arbitrary choice. Here a view has been selected by looking at the pictorial drawing from the bottom left-hand side, designated view C in the arrangement drawing in fig. 9.6. From this view, the other four possible views have been projected.

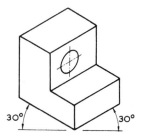

Fig. 9.5　Isometric pictorial projection of buffer stop

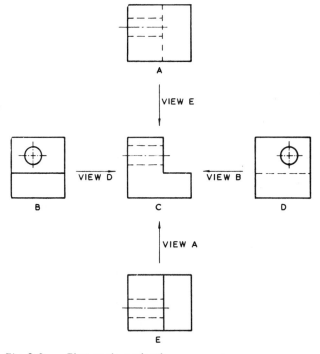

Fig. 9.6　First-angle projection

Each view is obtained by looking at the adjacent side of view C and projecting it onto the remote side, as indicated by the arrows. Alternatively, the component could be considered to be rolled over 90° in turn, in the direction of the arrows, and views A, B, D, and E would then face the draughtsman.

Normally for a detail drawing the draughtsman would select the minimum number of views to describe the part clearly, with the least amount of hidden detail; for the example given, views B and C would be acceptable. Only essential hidden detail should be shown by dotted lines.

Drawing in third-angle projection

Fig. 9.7 gives an alternative form of pictorial projection, to illustrate the principle of third-angle projection.

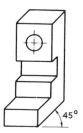

Fig. 9.7　Oblique pictorial projection example

In fig. 9.8, view C has again been chosen in an arbitrary manner by observing the pictorial component drawing from the left-hand side. Having selected this elevation, which is drawn first, the views A, B, D, and E are obtained by projection, in each case by looking at the adjacent side of view C.

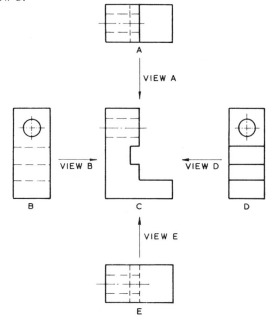

Fig. 9.8　Third-angle projection

In this example, views C and D are adequate to describe the component on a detail drawing.

In both figs 9.6 and 9.8, view C is known as the *front elevation*, views B and D are alternative *end elevations*, and views A and E are *plan views*.

Generally, industrial draughtsmen do not complete one view on a drawing before proceding to the next, but rather work on all views together. While projecting features between views, a certain amount of mental checking takes place regarding shape and form, and this assists accuracy. The following series of drawings shows stages in producing a typical working drawing in first-angle projection.

Stage 1 (fig. 9.9). Estimate the space required for each of the views from the overall dimensions in each plane, and position the views on the available drawing sheet so that the spaces between the three drawings are roughly the same.

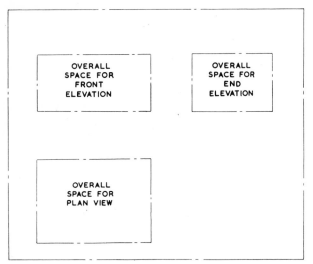

Fig. 9.9   Stage 1

Stage 2 (fig. 9.10). In each view, mark out the main centre-lines. Position any complete circles in any view, and line them in from the start, if possible. Here complete circles exist only in the plan view. The heights of the cylindrical features are now measured in the front elevation and are projected over to the end elevation.

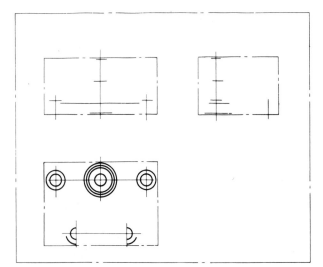

Fig. 9.10   Stage 2

Stage 3 (fig. 9.11). Complete the plan view and project up into the front elevation the sides of the cylindrical parts.

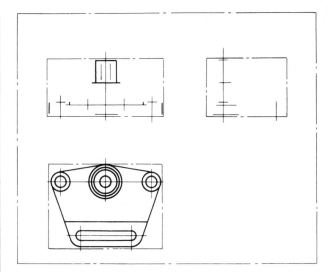

Fig. 9.11   Stage 3

Stage 4 (fig. 9.12). Complete the front and end elevations. Add dimensions, and check that the drawing (mental check) can be redrawn from the dimensions given; otherwise the dimensioning is incomplete. Add the title and any necessary notes.

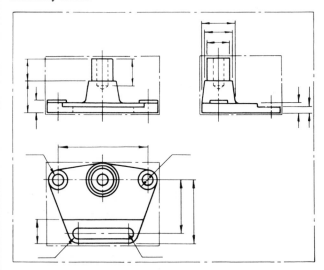

Fig. 9.12   Stage 4

It is generally advisable to mark out the same feature in as many views as is possible at the same time. Not only is this practice time-saving, but a continuous check on the correct projection between each view is possible, as the draughtsman then tends naturally to think in the three dimensions of length, breadth, and depth. It is rarely advantageous to complete one view before starting the others.

Chapter 10

Isometric and oblique projection

Isometric projection

Fig. 10.1 shows three views of a cube in orthographic projection; the phantom line indicates the original position of the cube, and the full line indicates the position after rotation about the diagonal AB. The cube has been rotated so that the angle of 45° between side AC_1 and diagonal AB now appears to be 30° in the front elevation, C_1 having been rotated to position C. It can clearly be seen in the end view that to obtain this result the angle of rotation is greater than 30°. Also, note that, although DF in the front elevation appears to be vertical, a cross check with the end elevation will confirm that the line slopes, and that point F lies to the rear of point D. However, the front elevation now shows a three-dimensional view, and when taken in isolation it is known as an *isometric projection*.

This type of view is commonly used in pictorial presentations, for example in car and motor-cycle service manuals and model kits, where an assembly has been 'exploded' to indicate the correct order and position of the component parts.

It will be noted that, in the isometric cube, line AC_1 is drawn as line AC, and the length of the line is reduced. Fig. 10.2 shows an isometric scale which in principle is obtained from lines at 45° and 30° to a horizontal axis. The 45° line

XY is calibrated in millimetres commencing from point X, and the dimensions are projected vertically onto the line XZ. By similar triangles, all dimensions are reduced by the same amount, and isometric lengths can be measured from point X when required. The reduction in length is in the ratio

$$\frac{\text{isometric length}}{\text{true length}} = \frac{\cos 45°}{\cos 30°} = \frac{0.7071}{0.8660} = 0.8165$$

Now, to reduce the length of each line by the use of an isometric scale is an interesting academic exercise, but commercially an isometric projection would be drawn using the true dimensions and would then be enlarged or reduced to the size required by photographic means.

Note that, in the isometric projection, lines AE and DB are equal in length to line AD; hence an equal reduction in length takes place along the apparent vertical and the two axes at 30° to the horizontal. Note also that the length of the diagonal AB does not change from orthographic to isometric, but that of diagonal C_1D_1 clearly does. When setting out an isometric projection, therefore, measurements must be made only along the isometric axes EF, DF, and GF.

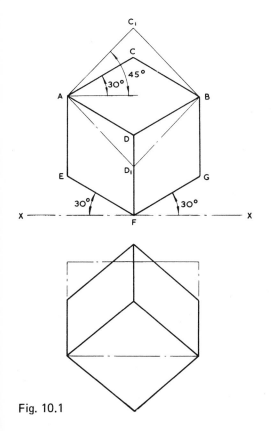

Fig. 10.1

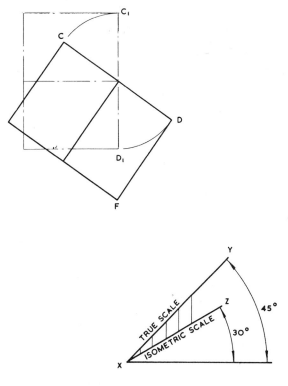

Fig. 10.2   Isometric scale

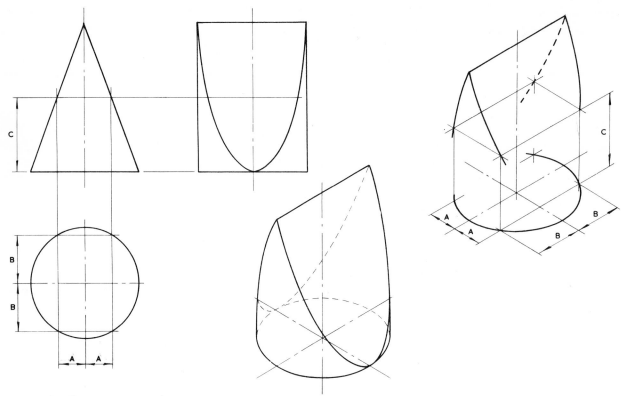

Fig. 10.3   Construction principles for points in space, with complete solution

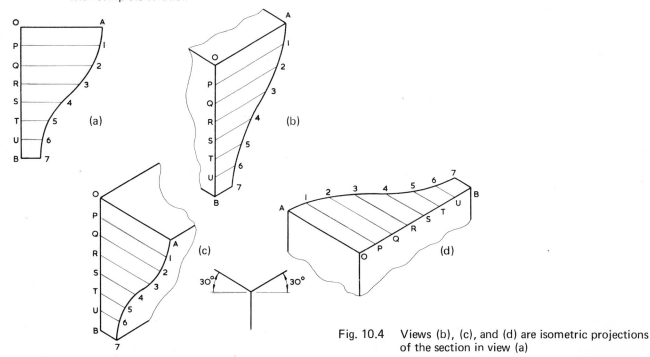

Fig. 10.4   Views (b), (c), and (d) are isometric projections of the section in view (a)

Fig. 10.3 shows a wedge which has been produced from a solid cylinder, and dimensions *A*, *B*, and *C* indicate typical measurements to be taken along the principle axes when setting out the isometric projection. Any curve can be produced by plotting a succession of points in space after taking ordinates from the *X, Y,* and *Z* axes.

Fig. 10.4 (a) shows a cross-section through an extruded alloy bar; the views (b), (c), and (d) give alternative isometric presentations drawn in the three principal planes of projection. In every case, the lengths of ordinates OP, OQ, P1, and Q2, etc. are the same but are positioned either vertically or inclined at 30° to the horizontal.

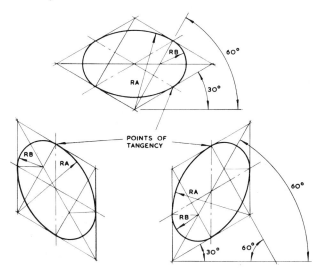

Fig. 10.5   Isometric circle construction

Fig. 10.5 shows an approximate method for the construction of isometric circles in each of the three major planes. Note the position of the points of intersection of radii R_A and R_B, which in each case are shown by a small dash normal to the profile.

The construction shown in fig. 10.5 can be used partly for producing corner radii. Fig. 10.6 shows a small block with radiused corners together with isometric projection which emphasises the construction to find the centres for the corner radii; this should be the first part of the drawing to be attempted. The thickness of the block is obtained from projecting back these radii a distance equal to the block thickness and at 30°. Line in those parts of the corners visible behind the front face, and complete the pictorial view by adding the connecting straight lines for the outside of the profile.

In the approximate construction shown, a small inaccuracy occurs along the major axis of the ellipse, and fig. 10.7 shows the extent of the error in conjunction with a plotted circle. In the vast majority of applications where

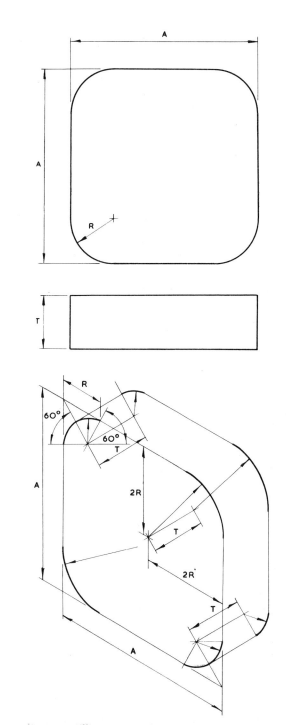

Fig. 10.6   Isometric constructions for corner radii

complete but small circles are used, for example spindles, pins, parts of nuts, bolts, and fixing holes, this error is of little importance and can be neglected.

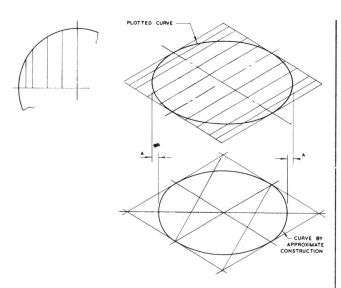

PLOTTED CURVE

CURVE BY APPROXIMATE CONSTRUCTION

Fig. 10.7   Relationship between plotted points and constructed isometric circles

Oblique projection

Fig. 10.8 shows part of a plain bearing in orthographic projection, and figs. 10.9 and 10.10 show alternative pictorial projections.

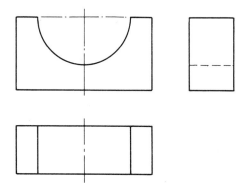

Fig. 10.8

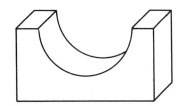

Fig. 10.9

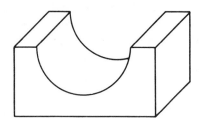

Fig. 10.10

It will be noted in figs 10.9 and 10.10 that the thickness of the bearing has been shown by projecting lines at 45° back from a front elevation of the bearing. Now, fig. 10.10 conveys the impression that the bearing is thicker than the true plan suggests, and therefore in fig. 10.9 the thickness has been reduced to one half of the actual size. Fig. 10.9 is known as an *oblique projection*, and objects which have curves in them are easiest to draw if they are turned, if possible, so that the curves are presented in the front elevation. If this proves impossible or undesirable, then fig. 10.11 shows part of the ellipse which results from projecting half sizes back along the lines inclined at 45°.

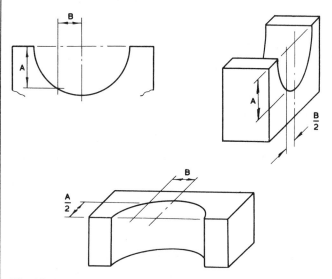

Fig. 10.11

A small die-cast lever is shown in fig. 10.12, to illustrate the use of a reference plane. Since the bosses are of different thicknesses, a reference plane has been taken along the side of the web; and, from this reference plane, measurements are taken forward to the boss faces and backwards to the opposite sides. Note that the points of tangency are marked, to position the slope of the web accurately.

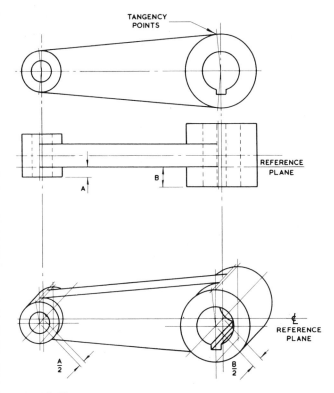

TANGENCY POINTS

REFERENCE PLANE

B

A

REFERENCE PLANE

$\frac{A}{2}$

$\frac{B}{2}$

Fig. 10.12

With oblique and isometric projections, no allowance is made for perspective, and this tends to give a slightly unrealistic result, since parallel lines moving back from the plane of the paper do not converge.

Chapter 11

Reading engineering drawings

Presentation of drawings in first- and third-angle projection has already been covered; the following notes and illustrations are intended to assist in the reading and understanding of simple drawings. In all orthographic drawings, it is necessary to project at least two views of a three-dimensional object—or one view and an adequate description in some simple cases, a typical example being the drawing of a ball for a bearing. A drawing of a circle on its own could be interpreted as the end elevation of a cylinder or a sphere. A drawing of a rectangle could be understood as part of a bar of rectangular cross-section, or it might be the front elevation of a cylinder. It is therefore generally necessary to produce at least two views, and these must be read together for a complete understanding. Fig. 11.1 shows various examples where the plan views are identical and the elevations are all different.

A single line may represent an edge or the change in direction of a surface, and which it is will be determined only by reading both views simultaneously. Fig. 11.2 shows other cases where the elevations are similar but the plan views are considerably different.

A certain amount of imagination is therefore required when interpreting engineering drawings. Obviously, with an object of greater complexity, the reading of three views, or more, may well be necessary.

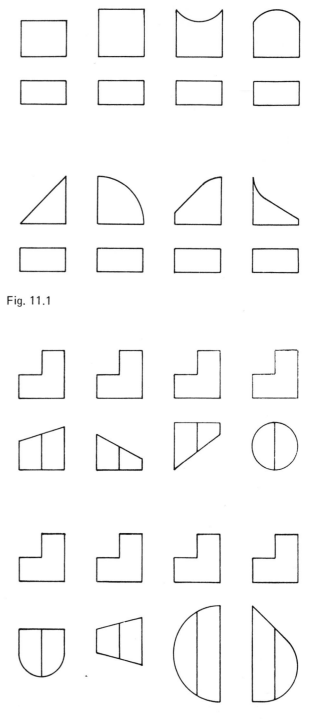

Fig. 11.1

Fig. 11.2

Chapter 12

Conic sections

Consider a right circular cone, i.e. a cone whose base is a circle and whose apex is above the centre of the base (fig. 12.1). The true face of a section through the apex of the cone will be a triangle.

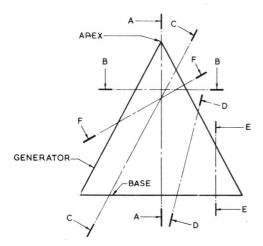

Fig. 12.1   Conic sections: section AA–triangle; section BB–circle; section CC–parabola; section DD–hyperbola; section EE–rectangular hyperbola; section FF–ellipse

The true face of a section drawn parallel to the base will be a circle.

The true face of any other section which passes through two opposite generators will be an ellipse.

The true face of a section drawn parallel to the generator will be a parabola.

If a plane cuts the cone through the generator and the base on the same side of the cone axis, then a view on the true face of the section will be a hyperbola. The special case of a section at right-angles to the base gives a rectangular hyperbola.

To draw an ellipse from part of a cone

Fig. 12.2 shows the method of drawing the ellipse, which is a true view on the surface marked AB of the frustum of the given cone.

1 Draw a centre line parallel to line AB as part of an auxiliary view.
2 Project points A and B onto this line and onto the centre lines of the plan and end elevation.
3 Take any horizontal section XX between A and B and draw a circle in the plan view of diameter D.
4 Project the line of section plane XX onto the end elevation.
5 Project the point of intersection of line AB and plane XX onto the plan view.

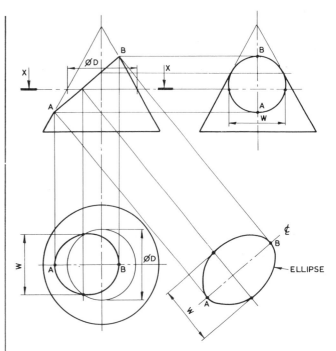

Fig. 12.2

6 Mark the chord-width W on the plan, in the auxiliary view and the end elevation. These points in the auxiliary view form part of the ellipse.
7 Repeat with further horizontal sections between A and B, to complete the views as shown.

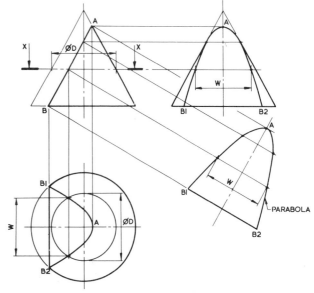

Fig. 12.3

To draw a parabola from part of a cone

Fig. 12.3 shows the method of drawing the parabola, which is a true view on the line AB drawn parallel to the sloping side of the cone.

1　Draw a centre line parallel to line AB as part of an auxiliary view.
2　Project point B to the circumference of the base in the plan view, to give the points B_1 and B_2. Mark chord-width $B_1 B_2$ in the auxiliary view and in the end elevation.
3　Project point A onto the other three views.
4　Take any horizontal section XX between A and B and draw a circle in the plan view of diameter D.
5　Project the line of section plane XX onto the end elevation.
6　Project the point of intersection of line AB and plane XX to the plan view.
7　Mark the chord-width W on the plan, in the end elevation and the auxiliary view. These points in the auxiliary view form part of the parabola.
8　Repeat with further horizontal sections between A and B, to complete the three views.

To draw a rectangular hyperbola from part of a cone

Fig. 12.4 shows the method of drawing the hyperbola, which is a true view on the line AB drawn parallel to the vertical centre line of the cone.

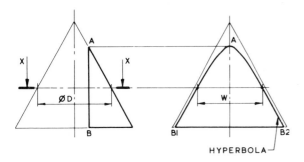

HYPERBOLA

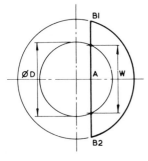

Fig. 12.4

1　Project point B to the circumference of the base in the plan view, to give the points B_1 and B_2.
2　Mark points B_1 and B_2 in the end elevation.
3　Project point A onto the end elevation. Point A lies on the centre line in the plan view.
4　Take any horizontal section XX between A and B and draw a circle of diameter D in the plan view.
5　Project the line of section XX onto the end elevation.
6　Mark the chord-width W in the plan, on the end elevation. These points in the end elevation form part of the hyperbola.
7　Repeat with further horizontal sections between A and B, to complete the hyperbola.

The ellipse, parabola, and hyperbola are also the loci of points which move in fixed ratios from a line (the directrix) and a point (the focus). The ratio is known as the *eccentricity*.

$$\text{Eccentricity} = \frac{\text{distance from focus}}{\text{perpendicular distance from directrix}}$$

The eccentricity for the ellipse is less than one.
The eccentricity for the parabola is one.
The eccentricity for the hyperbola is greater than one.

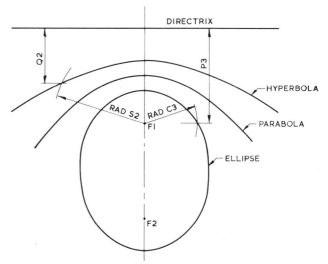

Fig. 12.5

Fig. 12.5 shows an ellipse of eccentricity 3/5, a parabola of eccentricity 1, and a hyperbola of eccentricity 5/3. The distances from the focus are all radial, and the distances from the directrix are perpendicular, as shown by the two illustrations.

To assist in the construction of the ellipse in fig. 12.5, the following method may be used to ensure that the two dimensions from the focus and directrix are in the same ratio. Draw triangle PA1 so that side A1 and side P1 are in the ratio of 3 units to 5 units. Extend both sides as shown. From any points B, C, D, etc., draw vertical lines to meet the horizontal at 2, 3, 4, etc.; by similar triangles, vertical lines and their corresponding horizontal lines will be in the same ratio. A similar construction for the hyperbola is shown in fig. 12.6.

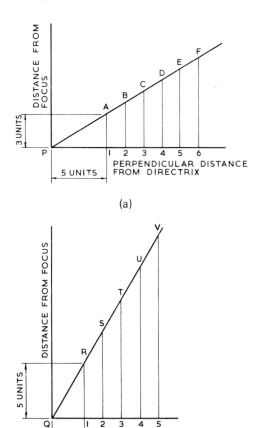

(a)

(b)

Fig. 12.6　(a) Ellipse construction　(b) Hyperbola construction

as the perpendicular distances and the radii are the same magnitude.

Repeat the procedure in each case to obtain the required curves.

Commence the construction for the ellipse by drawing a line parallel to the directrix at a perpendicular distance of P3 (fig. 12.6 (a)). Draw radius C3 from point F_1 to intersect this line. The point of intersection lies on the ellipse. Similarly, for the hyperbola (fig. 12.6 (b)) draw a line parallel to the directrix at a perpendicular distance of Q2. Draw radius S2, and the hyperbola passes through the point of intersection. No scale is required for the parabola,

Chapter 13

True lengths and auxiliary views

An isometric view of a rectangular block is shown in fig. 13.1. The corners of the block are used to position a line DF in space. Three orthographic views in first-angle projection are given in fig. 13.2, and it will be apparent that the projected length of the line DF in each of the views will be equal in length to the diagonals across each of the rectangular faces. A cross check with the isometric view will clearly show that the true length of line DF must be greater than any of the diagonals in the three orthographic views. The corners nearest to the viewing position are shown as ABCD etc.; the corners on the remote side are indicated in rings. To find the true length of DF, an auxiliary projection must be drawn, and the viewing position must be square with line DF. The first auxiliary projection in fig. 13.2 gives the true length required, which forms part of the right-angled triangle DFG. Note that auxiliary views are drawn on planes other than the principal projection planes. A plan is projected from an elevation and an elevation from a plan. Since this is the first auxiliary view projected, and from a true plan, it is known as a *first auxiliary elevation*. Other auxiliary views could be projected from this auxiliary elevation if so required.

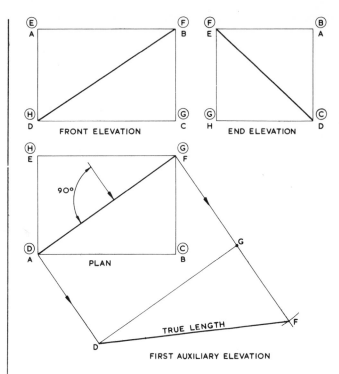

Fig. 13.2

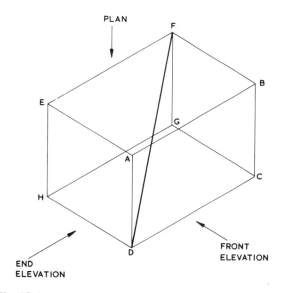

Fig. 13.1

The true length of DF could also have been obtained by projection from the front or end elevations by viewing at 90° to the line, and fig. 13.3 shows these two alternatives. The first auxiliary plan from the front elevation gives triangle FDH, and the first auxiliary plan from the end elevation gives triangle FCD, both right-angled triangles.

Fig. 13.4 shows the front elevation and plan view of a box. A first auxiliary plan is drawn in the direction of arrow X. Now PQ is an imaginary datum plane at right

angles to the direction of viewing; the perpendicular distance from corner A to the plane is shown as dimension 1. When the first auxiliary plan view is drawn, the box is in effect turned through 90° in the direction of arrow X, and the corner A will be situated above the plane at a perpendicular distance equal to dimension 1. The auxiliary plan view is a true view on the tilted box. If a view is now taken in the direction of arrow Y, the tilted box will be turned through 90° in the direction of the arrow, and dimension 1 to the corner will lie parallel with the plane of the paper. The other seven corners of the box are projected as indicated, and are positioned by the dimensions to the plane PQ in the front elevation. A match-box can be used here as a model to appreciate the position in space for each projection.

The same box has been redrawn in fig. 13.5, but the first auxiliary elevation has been taken from the plan view in a manner similar to that described in the previous example. The second auxiliary plan projected in line with arrow Y requires dimensions from plane P_1Q_1, which are taken as before from plane PQ. Again, check the projections shown with a match-box. All of the following examples use the principles demonstrated in these two problems.

Part of a square pyramid is shown in fig. 13.6; the constructions for the eight corners in both auxiliary views are identical with those described for the box in fig. 13.4.

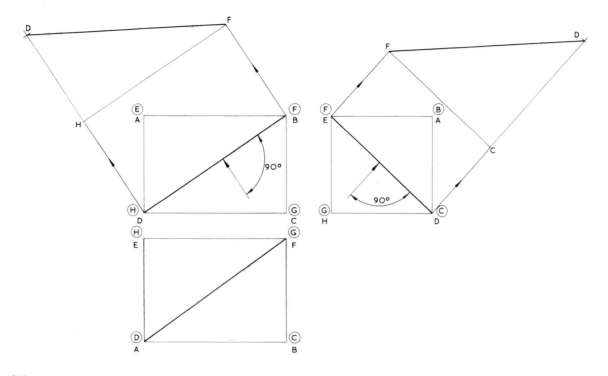

Fig. 13.3

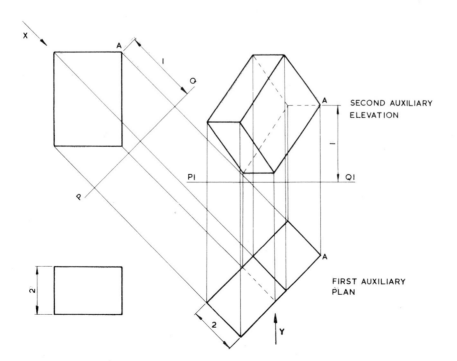

Fig. 13.4

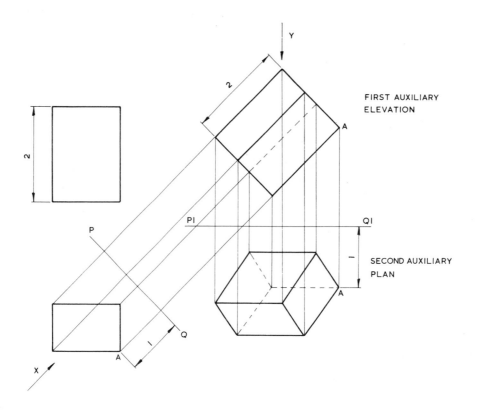

FIRST AUXILIARY
ELEVATION

SECOND AUXILIARY
PLAN

Fig. 13.5

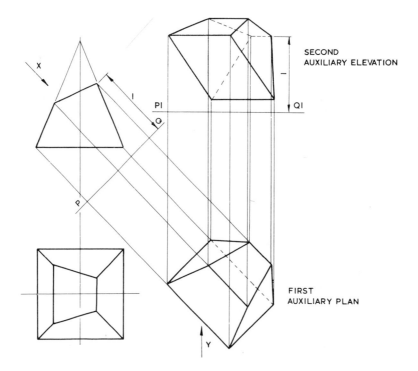

SECOND
AUXILIARY ELEVATION

FIRST
AUXILIARY PLAN

Fig. 13.6

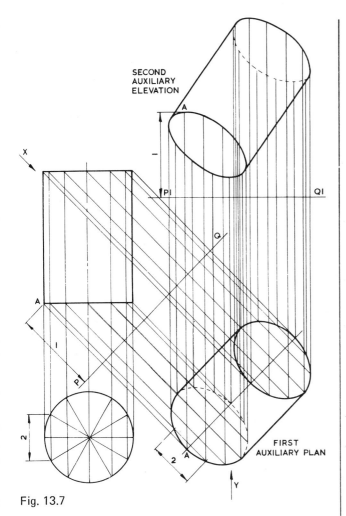

SECOND
AUXILIARY
ELEVATION

X

FIRST
AUXILIARY
PLAN

Fig. 13.7

Auxiliary projections from a cylinder are shown in fig. 13.7; note that chordal widths in the first auxiliary plan are taken from the true plan. Each of twelve points around the circle is plotted in this way and then projected up to the auxiliary elevation. Distances from plane PQ are used from plane P_1Q_1. Auxiliary projections of any irregular curve can be made by plotting the positions of a succession of points from the true view and rejoining them with a curve in the auxiliary view.

Fig. 13.8 shows a front elevation and plan view of a thin lamina in the shape of the letter L. The lamina lies inclined above the datum plane PQ, and the front elevation appears as a straight line. The true shape is projected above as a first auxiliary view. From the given plan view, an auxiliary elevation has been projected in line with the arrow F, and the positions of the corners above the datum plane P_1Q_1 will be the same as those above the original plane PQ. A typical dimension to the corner A has been added as

dimension 1. To assist in comprehension, the true shape given could be cut from a piece of paper and positioned above the book to appreciate how the lamina is situated in space; it will then be seen that the height above the book of corner A will be dimension 2.

Now a view in the direction of arrow G parallel with the surface of the book will give the lamina shown projected above datum P_2Q_2. The object of this exercise is to show that if only two auxiliary projections are given in isolation, it is possible to draw projections to find the true shape of the component and also get the component back, parallel to the plane of the paper. The view in direction of arrow H has been drawn and taken at 90° to the bottom edge containing corner A; the resulting view is the straight line of true length positioned below the datum plane P_3Q_3. The lamina is situated in this view in the perpendicular position above the paper, with the lower edge parallel to the paper and at a distance equal to dimension 4 from the surface. View J is now drawn square to this projected view and positioned above the datum P_4Q_4 to give the true shape of the given lamina.

In fig. 13.9, a lamina has been made from the polygon ACBD in the development and bent along the axis AB; again, a piece of paper cut to this shape and bent to the angle ϕ may be of some assistance. The given front elevation and plan position the bent lamina in space, and this exercise is given here since every line used to form the lamina in these two views is not a true length. It will be seen that, if a view is now drawn in the direction of arrow X, which is at right angles to the bend line AB, the resulting projection will give the true length of AB, and this line will also lie parallel with the plane of the paper. By looking along the fold in the direction of arrow Y, the two corners A and B will appear coincident; also, AD and BC will appear as the true lengths of the altitudes DE and FC. The development can now be drawn, since the positions of points E and F are known along the true length of AB. The lengths of the sides AD, DB, BC, and AC are obtained from the pattern development.

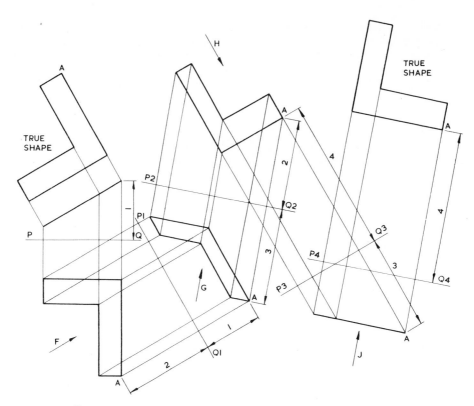

Fig. 13.8

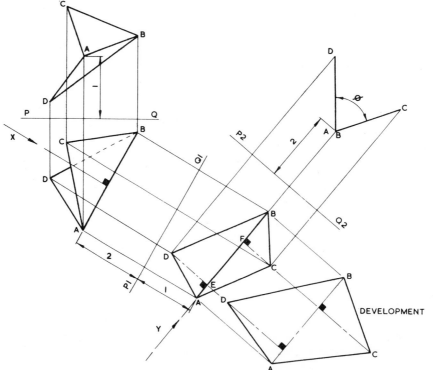

Fig. 13.9

Chapter 14

Interpenetration

Many objects are formed by a collection of geometrical shapes such as cubes, cones, spheres, cylinders, prisms, pyramids, etc., and where any two of these shapes meet, some sort of curve of intersection or interpenetration results. It is necessary to be able to draw these curves to complete drawings in orthographic projection or to draw patterns and developments.

The following drawings show some of the most commonly found examples of interpenetration. Basically, most curves are constructed by taking sections through the intersecting shapes, and, to keep construction lines to a minimum and hence avoid confusion, only one or two sections have been taken in arbitrary positions to show the principle involved; further similar parallel sections are then required to establish the line of the curve in its complete form. Where centre lines are offset, hidden curves will not be the same as curves directly facing the draughtsman, but the draughting principle of taking sections in the manner indicated on either side of the centre lines of the shapes involved will certainly be the same.

If two cylinders, or a cone and a cylinder, or two cones intersect each other at any angle, and the curved surfaces of both solids enclose the same sphere, then the outline of the intersection in each case will be an ellipse. In the illustrations given in fig. 14.1, the centre lines of the two solids intersect at point O, and a true view along the line AB will produce an ellipse.

When cylinders of equal diameter intersect as shown in fig. 14.2, the line at the intersection is straight and at 45°.

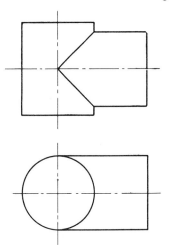

Fig. 14.2

Fig. 14.3 shows a branch cylinder square with the axis of the vertical cylinder but reduced in size. A section through any cylinder parallel with the axis produces a rectangle, in this case of width Y in the branch and width X in the vertical cylinder. Note that interpenetration occurs at points marked 3, and these points lie on a curve. The projection of the branch cylinder along the horizontal centre line gives the points marked 1, and along the vertical centre line gives the points marked 2.

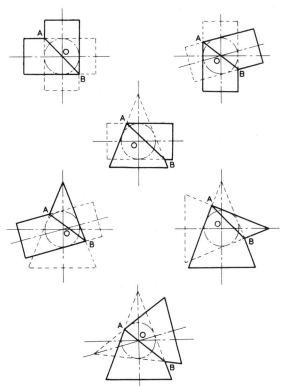

Fig. 14.1

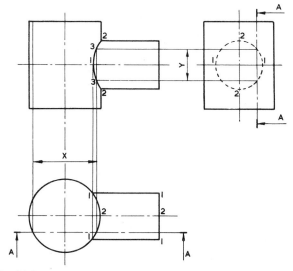

Fig. 14.3

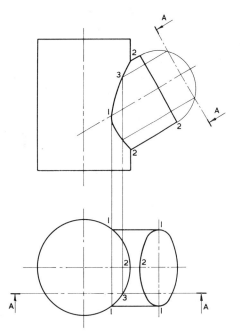

Fig. 14.4

Fig. 14.4 shows a cylinder with a branch on the same vertical centre line but inclined at an angle. Instead of an end elevation, the position of section AA is shown on a part auxiliary view of the branch. The construction is otherwise the same as that for fig. 14.3.

In fig. 14.5 the branch is offset, but the construction is similar to that shown in fig. 14.4

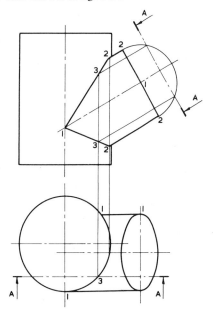

Fig. 14.5

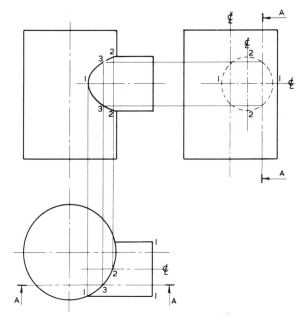

Fig. 14.6

Fig. 14.6 shows the branch offset but square with the vertical axis.

Fig. 14.7 shows a cone passing through a cylinder. A horizontal section AA through the cone will give a circle of ϕP, and through the cylinder will give a rectangle of width X. The points of intersection of the circle and part of the rectangle in the plan view are projected up to the section plane in the front elevation. The plotting of more points from more sections will give the interpenetration curves shown in the front elevation and the plan.

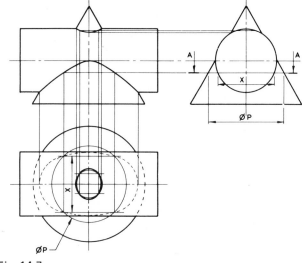

Fig. 14.7

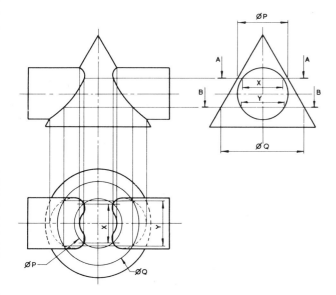

Fig. 14.8

Fig. 14.8 shows a cylinder passing through a cone. The construction shown is the same as for fig. 14.7 in principle.

Fig. 14.9 shows a cone and a square prism where interpenetration starts along the horizontal section BB at point 1 on the smallest diameter circle to touch the prism. Section AA is an arbitrary section where the projected diameter of the cone $\emptyset X$ cuts the prism in the plan view at the points marked 2. These points are then projected back to the section plane in the front elevation and lie on the curve required. The circle at section CC is the largest circle which will touch the prism across the diagonals in the plan

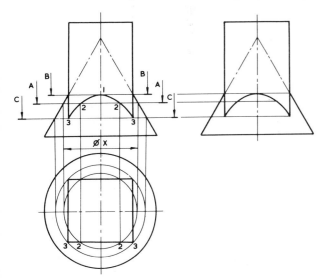

Fig. 14.9

view. Having drawn the circle in the plan view, it is projected up to the sides of the cone in the front elevation, and points 3 at the corners of the prism are the lowest points of contact.

A casting with a rectangular base and a circular-section shaft is given in fig. 14.10. The machining of the radius $R1$ in conjunction with the milling of the flat surfaces produces the curve shown in the front elevation. Point 1 is shown projected from the end elevation. Section AA produces a circle of $\emptyset X$ in the plan view and cuts the face of the casting at points marked 2, which are transferred back to the section plane. Similarly, section BB gives $\emptyset Y$ and points marked 3. Sections can be taken until the circle in the plan view increases in size to $R2$; at this point, the interpenetration curve joins a horizontal line to the corner of the casting in the front elevation.

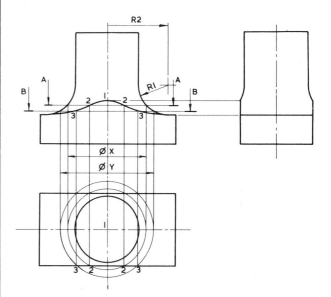

Fig. 14.10

In fig. 14.11 a circular bar of diameter D has been turned about the centre line CC and machined with a radius shown as RAD A. The resulting interpenetration curve is obtained by taking sections similar to section XX. At this section plane, a circle of radius B is projected in the front elevation and cuts the circumference of the bar at points E and F. The projection of point F along the section plane XX is one point on the curve. By taking a succession of sections, and repeating the process described, the curve can be plotted.

Note that, in all these types of problem, it rarely helps to take dozens of sections and then draw all the circles before plotting the points, as the only result is possible confusion. It is recommended that one section be taken at a

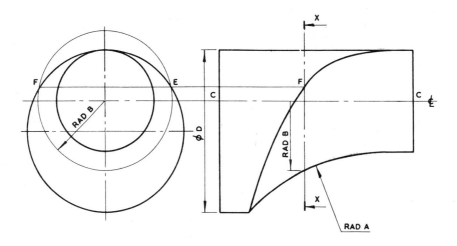

Fig. 14.11

time, the first roughly near the centre of any curve, and others sufficiently far apart for clarity but near enough to maintain accuracy. More sections are generally required where curves suddenly change direction.

Chapter 15

Developments

Many articles such as cans, pipes, elbows, boxes, ducting, hoppers, etc. are manufactured from thin sheet materials. Generally a template is produced from an orthographic drawing when small quantities are required (larger quantities may justify the use of press tools), and the template will include allowances for bending and seams, bearing in mind the thickness of material used.

Exposed edges which may be dangerous can be wired or folded, and these processes also give added strength, e.g. cooking tins and pans. Some cooking tins are also formed by pressing hollows into a flat sheet. This type of deformation is not considered in this chapter, which deals with bending or forming in one plane only. Some common methods of finishing edges, seams, and corners are shown in fig. 15.1.

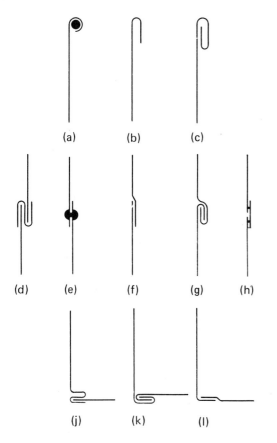

Fig. 15.1   Common methods of finishing edges, seams, and corners. (a) Wired edge (b) Single hem (c) Double hem (d) 'S'-type slip joint (e) Single-lap riveted joint (f) Single-lap soldered joint (g) Flat lock joint (h) Butt joint with spot-welded backing strip (j) Beaded joint (k) Corner cup joint (l) Single-lap soldered flush-corner joint

The following examples illustrate some of the more commonly used methods of development in pattern-making, but note that, apart from in the first case, no allowance has been made for joints and seams.

Where a component has its surfaces on flat planes of projection, and all the sides and corners shown are true lengths, the pattern is obtained by parallel-line or straight-line development. A simple application is given in fig. 15.2 for an open box.

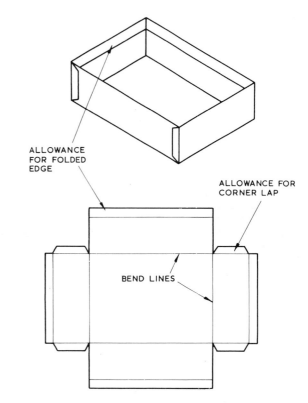

Fig. 15.2

The development of a hexagonal prism is shown in fig. 15.3. The pattern length is obtained by plotting the distances across the flat faces. The height at each corner is projected from the front elevation, and the top of the prism is drawn from a true view in the direction of arrow X.

An elbow joint is shown developed in fig. 15.4. The length of the circumference has been calculated and divided into twelve equal parts. A part plan, divided into six parts, has the division lines projected up to the joint, then across to the appropriate point on the pattern. It is normal practice on a development drawing to leave the joint along the shortest edge; however, on part B the pattern can be cut more economically if the joint on this half is turned through 180°.

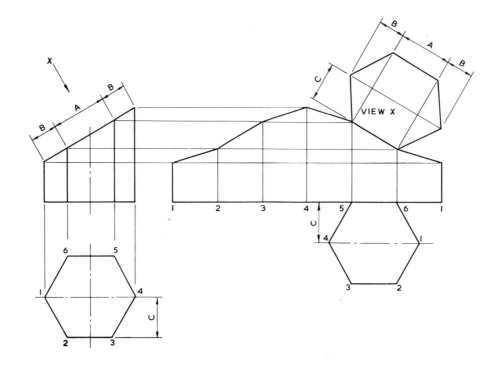

Fig. 15.3

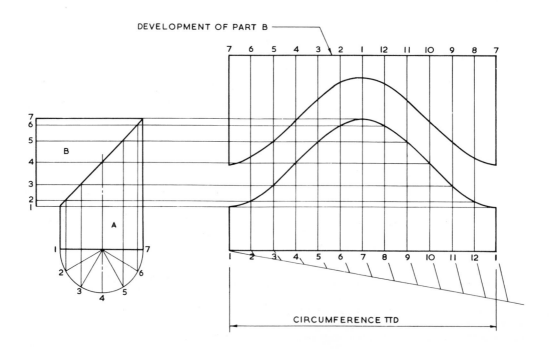

Fig. 15.4

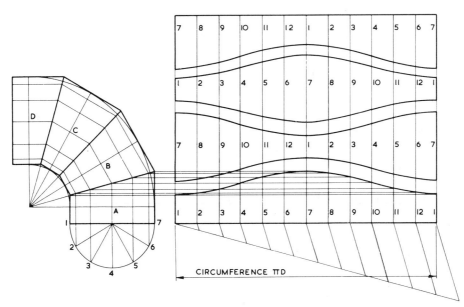

Fig. 15.5

An elbow joint made from four parts has been completely developed in fig. 15.5. Again, by alternating the position of the seams, the patterns can be cut with no waste. Note that the centre lines of the parts marked B and C are 30° apart, and that the inner and outer edges are tangential to the radii which position the elbow.

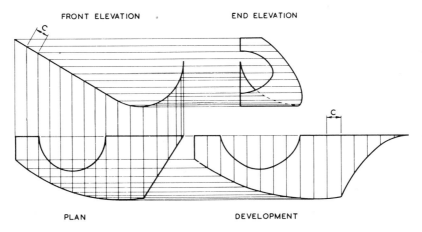

FRONT ELEVATION   END ELEVATION

PLAN   DEVELOPMENT

Fig. 15.6

A thin lamina is shown in orthographic projection in fig. 15.6. The development has been drawn in line with the plan view by taking the length along the front elevation in small increments of width *C* and plotting the corresponding depths from the plan.

A typical interpenetration curve is given in fig. 15.7. The development of part of the cylindrical portion is shown viewed from the inside. The chordal distances on the inverted plan have been plotted on either side of the centre line of the hole, and the corresponding heights have been projected from the front elevation. The method of drawing a pattern for the branch is identical to that shown for the two-piece elbow in fig. 15.4.

An example of radial-line development is given in fig. 15.8. The dimensions required to make the development are the circumference of the base and the slant height of the cone. The chordal distances from the plan view have been used to mark the length of arc required for the pattern; alternatively, for a higher degree of accuracy, the angle can be calculated and then subdivided. In the front elevation, lines O1 and O7 are true lengths, and distances OG and OA have been plotted directly onto the pattern. The lines O2 to O6 inclusive are not true lengths, and, where these lines cross the sloping face on the top of the conical frustum, horizontal lines have been projected to the side of the cone and been marked B, C, D, E, and F. True

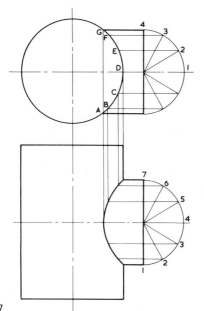

Fig. 15.7

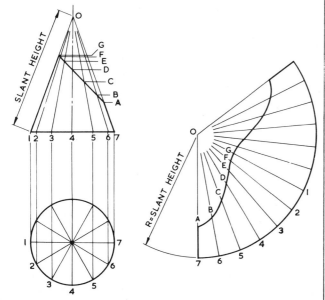

Fig. 15.8

lengths OF, OE, OD, OC, and OB are then marked on the pattern. This procedure is repeated for the other half of the cone. The view on the sloping face will be an ellipse, and the method of projection has been described in Chapter 12.

Part of a square pyramid is illustrated in fig. 15.9. The pattern is formed by drawing an arc of radius OA and stepping off around the curve the lengths of the base, joining the points obtained to the apex O. Distances OE

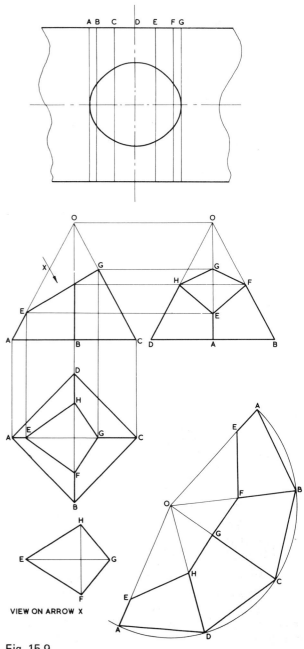

VIEW ON ARROW X

Fig. 15.9

and OG are true lengths from the front elevation, and distances OH and OF are true lengths from the end elevation. The true view in direction of arrow X completes the development.

The development of part of a hexagonal pyramid is shown in fig. 15.10. The method is very similar to that given in the previous example, but note that lines OB, OC, OD, OE, and OF are true lengths obtained by projection from the elevation.

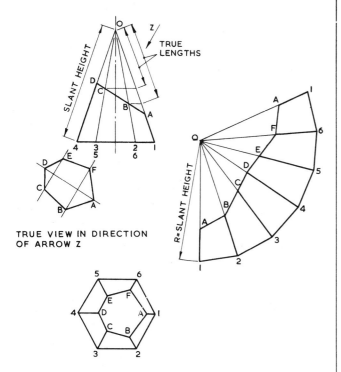

TRUE VIEW IN DIRECTION OF ARROW Z

Fig. 15.10

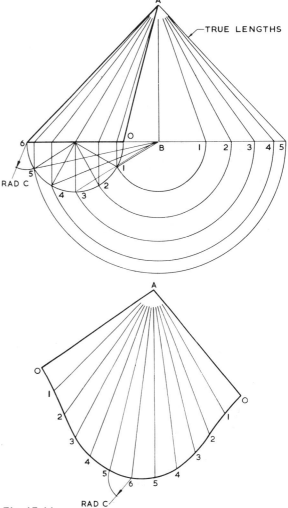

Fig. 15.11

Fig. 15.11 shows an oblique cone which is developed by triangulation, where the surface is assumed to be formed from a series of triangular shapes. The base of the cone is divided into a convenient number of parts (12 in this case) numbered 0-6 and projected to the front elevation with lines drawn up to the apex A. Lines 0A and 6A are true-length lines, but the other five shown all slope at an angle to the plane of the paper. The true lengths of lines 1A, 2A, 3A, 4A, and 5A are all equal to the hypotenuse of right-angled triangles where the height is the projection of the cone height and the base is obtained from the part plan view by projecting distances B1, B2, B3, B4, and B5 as indicated.

Assuming that the join will be made along the shortest edge, the pattern is formed as follows. Start by drawing line 6A, then from A draw an arc on either side of the line equal

in length to the true length 5A. From point 6 on the pattern, draw an arc equal to the chordal distance between successive points on the plan view. This curve will intersect the first arc twice at the points marked 5. Repeat by taking the true length of line 4A and swinging another arc from point A to intersect with chordal arcs from points 5. This process is continued as shown on the solution.

Fig. 15.12 shows the development of part of an oblique cone where the procedure described above is followed. The points of intersection of the top of the cone with lines 1A, 2A, 3A, 4A, and 5A are transferred to the appropriate true-length constructions, and true-length distances from the apex A are marked on the pattern drawing.

A plan and front elevation is given in fig. 15.13 of a transition piece which is formed from two halves of oblique cylinders and two connecting triangles. The plan view of

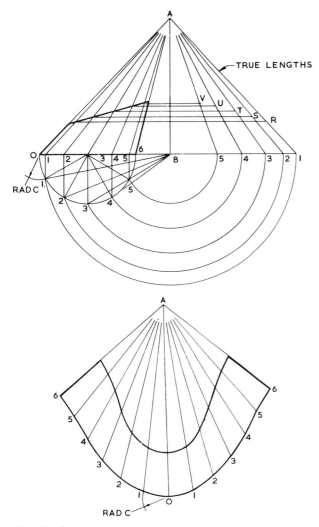

Fig. 15.12

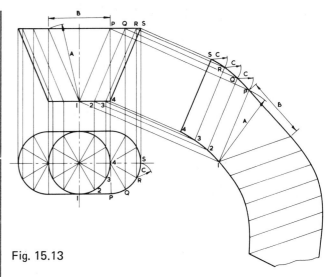

Fig. 15.13

Part of a triangular prism is shown in fig. 15.14, in orthographic projection. The sides of the prism are constructed from a circular arc of true radius OC in the end elevation. Note that radius OC is the only true length of a sloping side in any of the three views. The base length CA is

the base is divided into 12 equal divisions, the sides at the top into 6 parts each. Each division at the bottom of the front elevation is linked with a line to the similar division at the top. These lines, P1, Q2, etc., are all the same length. Commence the pattern construction by drawing line S4 parallel to the component. Project lines from points 3 and R, and let these lines intersect with arcs equal to the chordal distances *C*, from the plan view, taken from points 4 and S. Repeat the process and note the effect that curvature has on the distances between the lines projected from points P, Q, R, and S. After completing the pattern to line P1, the triangle is added by swinging an arc equal to the length *B* from point P, which intersects with the arc shown, radius *A*. This construction for part of the pattern is continued as indicated.

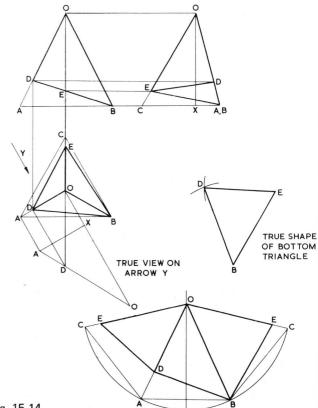

Fig. 15.14

marked around the circumference of the arc three times, to obtain points A, B, and C. True length OE can be taken from the end elevation, but a construction is required to find the true length of OD. Draw an auxiliary view in direction with arrow Y, which is square to line OA as shown. The height of the triangle, OX, can be taken from the end elevation. The projection of point D on the side of the triangle gives the true length OD. The true shape at the bottom can be drawn by taking lengths ED, DB, and BE from the pattern and constructing the triangle shown.

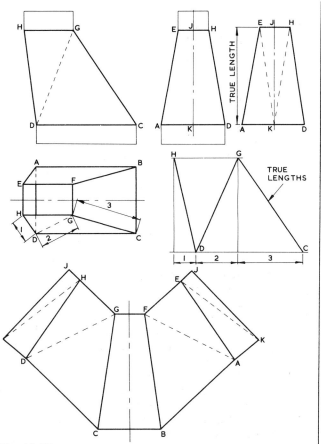

Fig. 15.15

A transition piece connecting two rectangular ducts is given in fig. 15.15. The development is commenced by drawing the figure CBFG, and the centre line of this part can be obtained from the front elevation which appears as line CG, the widths being taken from the plan. The next problem is to obtain the true lengths of lines CG and DH and position them on the pattern; this can be done easily by the construction of two triangles, after the insertion of line DG. The true lengths can be found by drawing right-angled triangles where the base measurements are

indicated as dimensions 1, 2, and 3, and the height is equal to the height of the front elevation. The length of the hypotenuse in each case is used as the radius of an arc to form triangles CDG and GDH. The connecting seam is taken along the centre line of figure ADHE and is marked JK. The true length of line JK appears as line HD in the front elevation, and the true shape of this end panel has been drawn beside the end elevation to establish the true lengths of the dotted lines EK and HK, since these are used on the pattern to draw triangles fixing the exact position of points K and J.

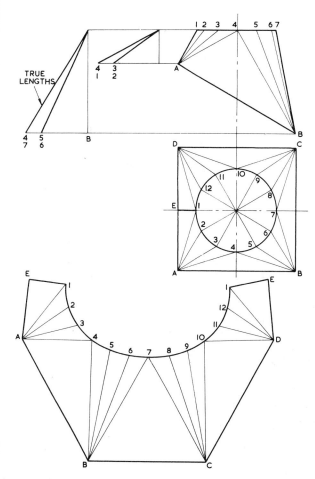

Fig. 15.17

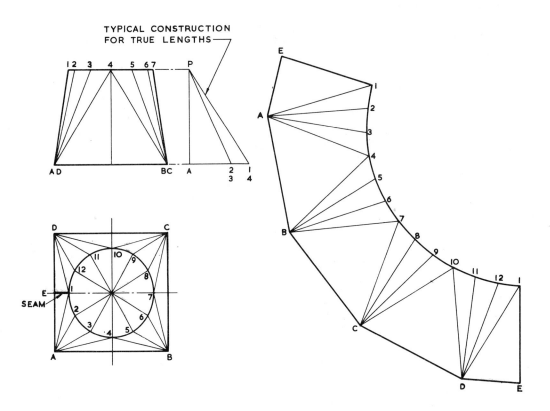

Fig. 15.16

A transition piece connecting square and circular ducts is shown in fig. 15.16. The circle is divided into twelve equal divisions, and triangles are formed on the surface of the component as shown. A construction is required to establish the true lengths of lines A1, A2, A3, and A4. These lengths are taken from the hypotenuse of right-angled triangles whose height is equal to the height of the front elevation, and the base measurement is taken from the projected lengths in the plan view. Note that the lengths A2 and A3 are the same, as are A1 and A4, since the circle lies at the centre of the square in the plan. The constructions from the other three corners are identical to those from corner A. To form the pattern, draw a line AB, and from A describe an arc of radius A4. Repeat from end B, and join the triangle. From point 4, swing an arc equal to the chordal length between points 4 and 3 in the plan view, and let this arc intersect with the true length A3, used as a

radius from point A. Mark the intersection as point 3. This process is repeated to form the pattern shown. The true length of the seam at point E can be measured from the front elevation. Note that, although chordal distances are struck between successive points around the pattern, the points are themselves joined by a curve; hence no ultimate error of any significance occurs when using this method.

Fig. 15.17 shows a similar transition piece where the top and bottom surfaces are not parallel. The construction is generally very much the same as described above, but two separate true-length constructions are required for the corners marked AD and BC. Note that, in the formation of the pattern, the true length of lines AB and CD is taken from the front elevation when triangles AB4 and DC10 are formed. The true length of the seam is also the same as line A1 in the front elevation.

Chapter 16

Types of drawings and layouts

Single-part drawing

A single-part drawing should supply the complete detailed information to enable a component to be manufactured without reference to other sources. It should completely define shape or form and size, and should contain a specification. The number of views required depends on the degree of complexity of the component. The drawing must be fully dimensioned, including tolerances where necessary, to show all sizes and locations of the various features. The specification for the part includes information relating to the material used and possible heat-treatment required, and notes regarding finish. The finish may apply to particular surfaces only, and may be obtained by using special machining operations or, for example, by plating, painting, or enamelling. Fig. 16.1 shows typical single-part drawings.

An alternative to a single-part drawing is to collect several small details relating to the same assembly and group them together on the same drawing sheet. In practice, grouping in this manner may be satisfactory provided all the parts are made in the same department, but it can be inconvenient where, for example, pressed parts are drawn with turned components or sheet-metal fabrications.

More than one drawing may also be made for the same component. Consider a sand-cast bracket. Before the bracket is machined, it needs to be cast; and, before casting, a pattern needs to be produced by a patternmaker. It may therefore be desirable to produce for the patternmaker a drawing which includes the various machining allowances, and then produce a separate drawing for the benefit of the machinist which shows only dimensions relating to the surfaces to be machined and the size of the finished part. The two drawings would each have only parts of the specification which suited that particular manufacturing process.

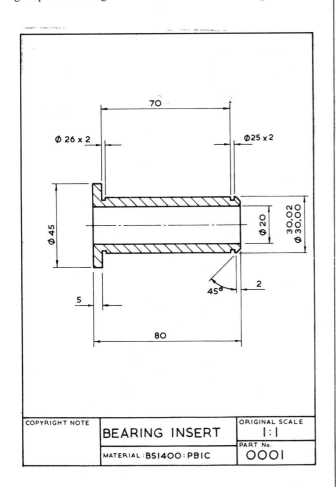

Fig. 16.1 (a) Bearing insert

Fig. 16.1 (b) Gear hub

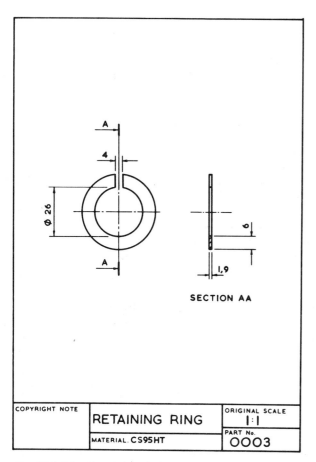

Fig. 16.1 (c)   Retaining ring

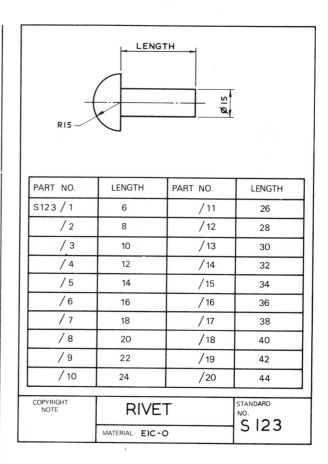

Fig. 16.2   Collective single-part drawing of a rivet

Collective single-part drawings

Fig. 16.2 shows a typical collective single-part drawing for a rivet. The drawing covers 20 rivets similar in every respect except length; in the example given, the part number for a 30mm rivet is S123/13. This type of drawing can also be used where, for example, one or two dimensions on a component (which are referred to on the drawing as *A* and *B*) are variable, all others being standard. For a particular application, the draughtsman would insert the appropriate value of dimensions *A* and *B* in a table, then add a new suffix to the part number. This type of drawing can generally be used for basically similar parts.

Assembly drawings

Machines and mechanisms consist of numerous parts, and a drawing which shows the complete product with all its components in their correct physical relationship is known as an assembly drawing. A drawing which gives a small part of the whole assembly is known as a sub-assembly drawing. A sub-assembly may in fact be a complete unit itself; for example, a drawing of a clutch could be considered as a sub-assembly of a drawing showing a complete automobile engine. The amount of information given on an assembly drawing will vary considerably with the product and its size and complexity.

If the assembly is relatively small, information which might be given includes a parts list. The parts list, as the name suggests, lists the components, which are numbered. Numbers in 'balloons' with leader lines indicate the position of the component on the drawing—see fig. 16.3. The parts list will also contain information regarding the number required of each component for the assembly, its individual single-part drawing number, and possibly its material. Parts lists are not standard items, and their contents vary from one drawing office to another.

The assembly drawing may also give other information, including overall dimensions of size, details of bolt sizes and centres where fixings are necessary, weights required for shipping purposes, operating details and instructions, and

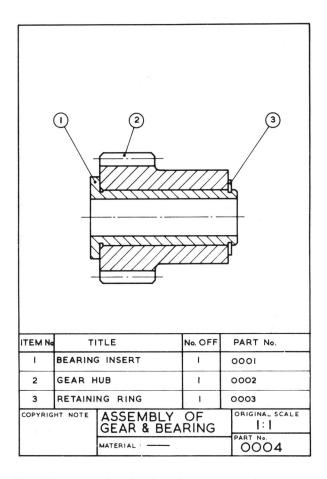

Fig. 16.3   Assembly drawing of gear and bearing

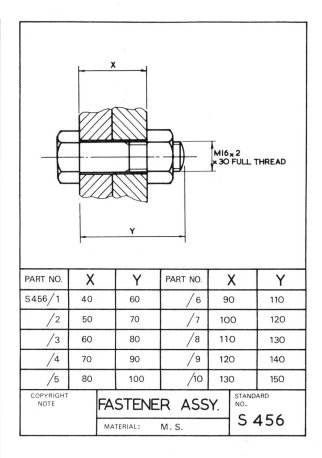

PART NO.	X	Y	PART NO.	X	Y
S456/1	40	60	/6	90	110
/2	50	70	/7	100	120
/3	60	80	/8	110	130
/4	70	90	/9	120	140
/5	80	100	/10	130	150

Fig. 16.4   Typical collective assembly drawing of a nut
with bolts of various lengths.

also, perhaps, some data regarding the design characteristics.

Collective assembly drawing

This type of drawing is used where a range of products which are similar in appearance but differing in size is manufactured and assembled. Fig. 16.4 shows a nut-and-bolt fastening used to secure plates of different combined thickness; the nut is standard, but the bolts are of different lengths. The accompanying table is used to relate the various assemblies with different part numbers.

Combined detail and assembly drawings

It is sometimes convenient to illustrate details with their assembly drawing on the same sheet. This practice is particularly suited to small 'one-off' or limited-production-run assemblies. It not only reduces the actual number of

drawings, but also the drawing-office time spent in scheduling and printing. Fig. 16.5 shows a simple application of an assembly of this type.

Exploded assembly drawings

Fig. 16.6 shows a typical exploded assembly drawing from a model kit; these drawings are prepared to assist in the correct understanding of the various component positions in an assembly. Generally a pictorial type of projection is used, so that each part will be shown in three dimensions. Exploded views are invaluable when undertaking servicing and maintenance work on all forms of plant and appliances. Car manuals and do-it-yourself assembly kits use such drawings, and these are easily understood. As well as an aid to construction, an exploded assembly drawing suitably numbered can also be of assistance in the ordering of spare parts; components are more easily recognisable in a pictorial projection, especially by people without training in the reading of technical drawings.

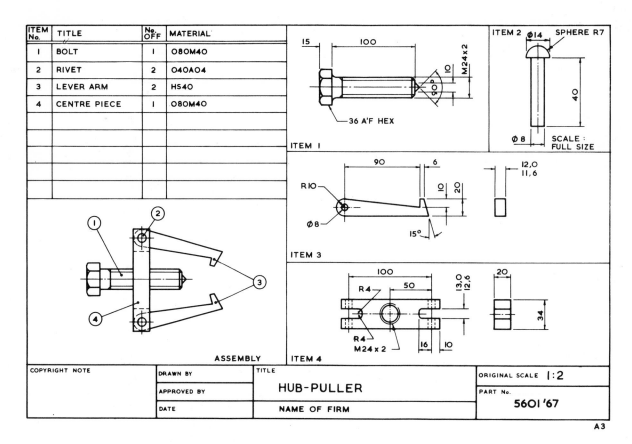

ITEM No.	TITLE	No. OFF	MATERIAL
1	BOLT	1	080M40
2	RIVET	2	040A04
3	LEVER ARM	2	HS40
4	CENTRE PIECE	1	080M40

ITEM 1

ITEM 2   SCALE: FULL SIZE

ITEM 3

ASSEMBLY   ITEM 4

COPYRIGHT NOTE	DRAWN BY	TITLE	ORIGINAL SCALE 1:2
	APPROVED BY	HUB-PULLER	PART No.
	DATE	NAME OF FIRM	5601 '67

A3

Fig. 16.5 Combined detail and assembly drawing of hub-puller

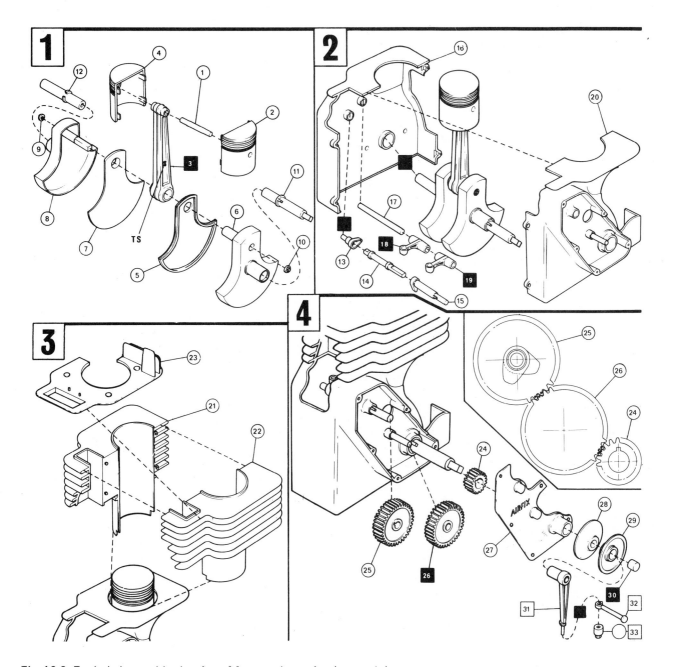

Fig. 16.6 Exploded assembly drawing of four-stroke engine (part only)

Chapter 17

Simplified drawings

Simplified draughting conventions have been devised to reduce the time spent drawing and detailing symmetrical components and repeated parts. Fig. 17.1 shows a gasket which is symmetrical about the horizontal centre line. A detail drawing indicating the line of symmetry and half of the gasket is shown in fig. 17.2, and this is sufficiently clear for the part to be manufactured.

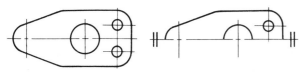

Fig. 17.1   Fig. 17.2

If both halves are similar except for a small detail, then the half which contains the exception is shown with an explanatory note to that effect, and a typical example is illustrated in fig. 17.3.

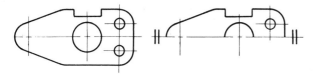

Fig. 17.3   When dimensioning, add drawing note 'SLOT ON ONE SIDE'

A joint-ring is shown in fig. 17.4, which is symmetrical about two axes of symmetry. Both axes are shown in the detail, and a quarter view of the joint-ring is sufficient for the part to be made.

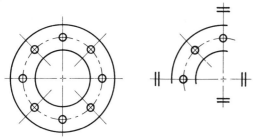

Fig. 17.4

The practice referred to above is not restricted to flat thin components, and fig. 17.5 gives a typical detail of a straight lever with a central pivot in part section. Half the lever is shown, since the component is symmetrical, and a partial view is added and drawn to an enlarged scale to clarify the shape of the boss and leave adequate space for dimensioning.

Repeated information also need not be drawn in full; for example, to detail the peg-board in fig. 17.6, all that is

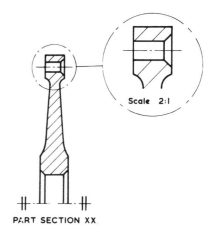

Fig. 17.5   Part of a lever detail drawing, symmetrical about the horizontal axis.

required is to draw one hole, quoting its size and fixing the centres of all the others.

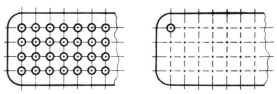

Fig. 17.6

Similarly, fig. 17.7 shows a gauze filter. Rather than draw the gauze over the complete surface area, only a small portion is sufficient to indicate the type of pattern required.

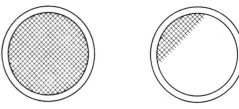

Fig. 17.7

Knurled screws are shown in fig. 17.8 to illustrate the accepted conventions for straight and diamond knurling.

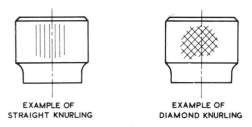

EXAMPLE OF STRAIGHT KNURLING   EXAMPLE OF DIAMOND KNURLING

Fig. 17.8

Chapter 18

Sections

A section is drawn to show the shape, form, or internal construction of a component or assembly where an outside view would prove inadequate. Consider the plastics pulley shown in fig. 18.1 and imagine that it is cut with a hacksaw along the line AA. The line AA is called the *cutting plane*, shown as a long-chain line 0.3mm thick and thickened at both ends. The cutting plane is lettered AA, and SECTION AA is written under the projected elevation. More involved

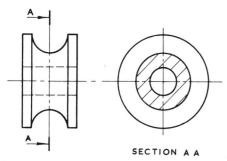

SECTION A A

Fig. 18.1

parts may have many cutting planes and projected sections to clearly illustrate involved details. Note that the arrows at each end of the plane are square with it and point to the right. Since we are drawing in first-angle projection, we discard the left-hand half of the pulley and project to the right the view in line with the arrows. The actual plastics cut through is shown cross-hatched. The hatching should be at 45° to the centre lines, with continuous lines 0.3 mm thick and successive lines not closer than 4mm apart; hatching lines drawn closer than 4mm tend to merge when microfilmed. The distance between hatching lines should be in proportion with the size of the component or components. Fig. 18.2 shows a bush in a casting; note that the hatching for the bush, which is the smaller part, is closer. The cross-hatching on the casting is not only wider but in

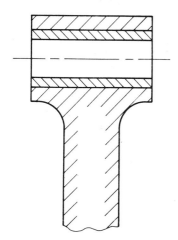

Fig. 18.2

the opposite direction, and this is customary on adjacent parts. The hatching over large areas can be restricted to the area close to the boundary, as shown in fig. 18.2, but much care needs to be taken here, otherwise drawings substandard in appearance can easily result. Hatching can be partially omitted so that notes or dimensions can be placed on a drawing, as shown in fig. 18.3.

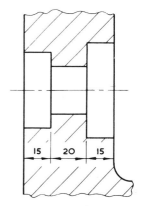

Fig. 18.3

Centre lines of components need not be horizontal or vertical, but hatching should be at 45° to the axis of the part, regardless of the angle at which the part lies, as shown in fig. 18.4.

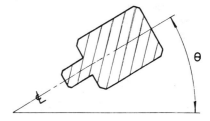

Fig. 18.4

Some sections are too thin to be cross-hatched, a typical example of a clip and support bracket in cross-section being given in fig. 18.5. The clip and bracket are drawn apart and are indicated by thick black lines. The same situation often applies with structural steel sections drawn to scale and sections through sheet-metal fabrications, gaskets, seals, and packings.

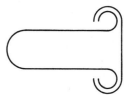

Fig. 18.5

The section shown in fig. 18.1 was a full section with a straight section plane. It is sometimes required to include in the same elevation details from two or more part sections from parallel planes; an example is given in fig. 18.6. Note that the section line is thickened where the section plane changes direction.

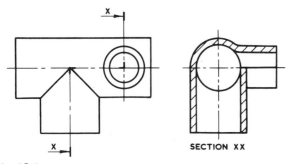

Fig. 18.6

Fig. 18.7 shows a sectioned elevation from a plan where the section line is taken along. three neighbouring planes which are not at right angles to one another. The section line follows the section planes in order, and is thickened at each change of direction.

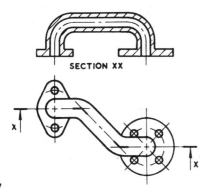

Fig. 18.7

Figs 18.8 and 18.9 show the same disc with two drilled holes in different positions. Fig. 18.8 gives a full section in one plane, and fig. 18.9 illustrates the application of a revolved section plane. The section plane changes in direction to incorporate a feature, and is revolved back into the vertical position so that the end elevation becomes a true projection on the section plane. Note that both arrows are still drawn square with the section lines. It will be seen that the end elevations in figs 18.8 and 18.9 are the same.

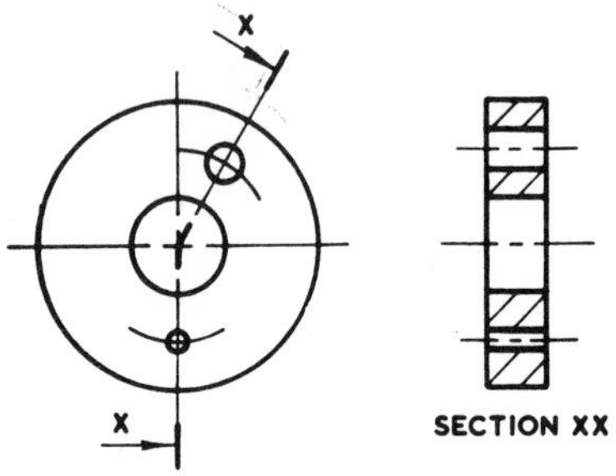

Fig. 18.10 shows a component which has grinding centres at both ends, a flat for a cotter pin, and four square notches on a shouldered portion. Although not dimensioned, successive sections are shown along the pin to illustrate the shape of the cross-sections at two points. The

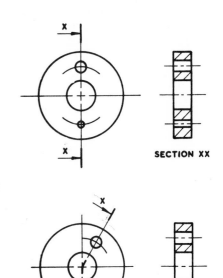

Fig. 18.8

Fig. 18.9

flat for the cotter pin is also indicated by diagonal lines. Note the use of a local part section to give details of the grinding centres.

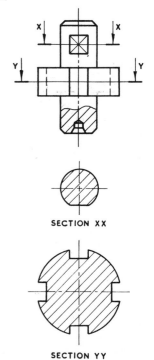

Fig. 18.10

The clarity of a single-view drawing is often improved if presented in half section as shown in fig. 18.11, where the left-hand half is a section along the centre line and the right-hand part is an outside view. This type of drawing avoids the necessity of introducing dotted lines for the holes and the recess. Dimensioning to dotted lines is not a recommended practice.

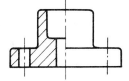

Fig. 18.11

Fig. 18.12 shows a method of presenting two sections from parallel planes along the same symmetrical part. To indicate the change of position of the section plane, the hatching is offset but sloping in the same direction.

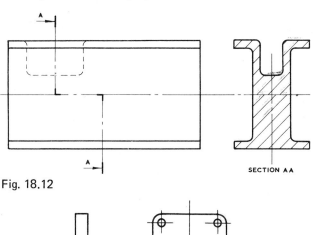

SECTION AA

Fig. 18.12

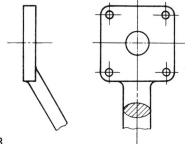

Fig. 18.13

Fig. 18.13 gives an example of a revolved section (which must be drawn with a 0.3mm line) on the auxiliary elevation. This method involves rotation of the section through 90° into the plane of the paper. An alternative method is indicated in fig. 18.14, where a removed section AA is drawn. Note that no additional background information has been included, since the removed section only indicates the true shape of the casting at the point where

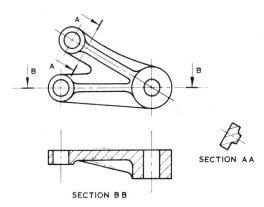

SECTION AA

SECTION BB

Fig. 18.14

the section has been taken. Section BB shows a section along the horizontal centre line through a thin web. It is not common practice to section thin webs, since this gives an impression of solidity and could prove misleading; however, no doubt can exist, because a second view would be present to show the thickness of the web. This section along the centre line should truly indicate the outline of the web in a dotted line. In no circumstances is cross-hatching ever taken up to dotted lines, so in this case the web is defined by a full line.

It is also not customary to section nuts, bolts, washers, rivets, keys, pins, dowels, shafts, or other similar solid components, except for local part sections as already indicated in fig. 18.10.

Chapter 19

Dimensioning

A drawing should provide a complete set of working instructions for the craftsmen who produce the component. Dimensions define geometric characteristics such as angles, diameters, lengths, and positions; and each dimension which defines a characteristic should appear on the drawing only once, and it should not be necessary for the craftsman either to scale the drawing or to deduce dimensions by the subtraction or addition of other dimensions.

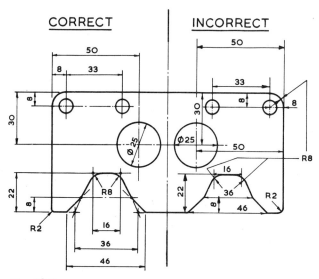

Fig. 19.1

Fig. 19.1 shows a component which has been partly dimensioned to illustrate some of the principles involved.

1. Projection and dimension lines are thin continuous lines drawn 0.3mm thick.
2. Where possible, dimensions should be placed outside the outline of the drawing, for clarity.
3. Projection lines 0.3mm thick should not touch the drawing, but should start at about 3mm from the feature and continue about 3mm past the dimension line. Intersection points may be emphasized by small dots, and, if these are used, the projection lines should pass through the dots.
4. Dimensions should not be cramped, but should be spaced about 12mm apart, the smaller dimensions nearer to the component and overall dimensions furthest from the outline. This practice avoids projection and dimension lines crossing. Note that a spacing of 12mm permits upper and lower limits of size to be inserted where applicable. Dimension lines may be broken to insert the figures if required, but it is preferable to show the figure near to the centre and just above the dimension line.
5. Arrow-heads should be uniform in size, about 4mm long and 2mm thick, and should touch the projection line in every case.

6. Centre lines must never be used as dimension lines.
7. To assist in the reading of dimensions, figures should be placed so that they can be read easily from the bottom of the sheet or, by turning the drawing in a clockwise direction, from the right-hand side of the drawing sheet.
8. Avoid long intersecting leader lines, as shown from the R8 dimension; it is clearer to repeat dimensions in such cases. Note that the dimension line to a radius either passes through the centre of the arc, or, if it is placed outside the outline, it is in line with the centre of the arc.
9. Dimensions are quoted in millimetres to the minimum number of significant figures, for example 22 and not 22.0. A decimal dimension would be quoted as, for example, 0.5 and not .5. It is possible with a dirty print in a workshop to loose sight of the decimal point, and the zero figure preceding the decimal dimension removes doubt.

On metric drawings, the decimal marker can be shown as either a full stop or a comma on the base line between figures—never midway. Whichever method is used on a drawing, it should be consistent, and under no circumstances should a combination of the two methods be used.

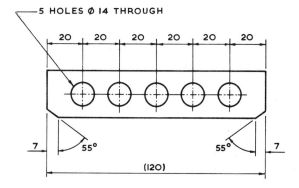

Fig. 19.2

Fig. 19.2 shows an example of chain dimensioning, with an auxiliary dimension added. The auxiliary dimension shown in brackets is not to be used for manufacture but is quoted only for guidance. Note that in this case each of the 20mm pitches would be subject to a manufacturing tolerance, and the overall length may vary considerably depending on the tolerance permitted. All of the pitches may well be marked out at the high limit of size, or alternatively at the low limit of size, and the length will clearly depend on the summation of all the individual distances. To avoid the possible accumulation of tolerances, two alternative methods are used and are shown in fig. 19.3. Here the

Method 1—standard method

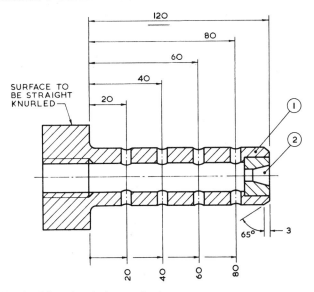

Method 2— alternative method

Fig. 19.3

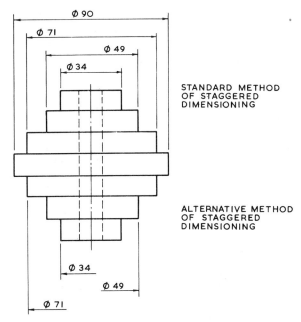

STANDARD METHOD
OF STAGGERED
DIMENSIONING

ALTERNATIVE METHOD
OF STAGGERED
DIMENSIONING

Fig. 19.4   Typical belt pulley

datum selected is under the head of the component. Method 1 shows the four holes dimensioned back to the datum, each dimension subject to a manufacturing tolerance which cannot accumulate to the detriment of the functional requirements of the component. Where drillings are to be indicated, for example, on long structural sections, and space would prohibit this style of dimensioning, then the second method can be used. The dimensions are placed in line, and the datum is indicated by a large dot. The figures are positioned close to the arrow-head in each case.

The two part numbers for the separate components are contained in circles, and leader lines indicate the parts in the assembly. Leader lines terminate in a dot within the outline of the parts. The leader line indicating the knurled surface terminates with an arrow on the surface. The angle between the leader line and the surface should preferably be drawn between 60°–80°. The chamfer at the end of the component is dimensioned by stating the angle and the depth of chamfer parallel to the axis. The dimension of 120 which has been underlined indicates that the length has been drawn out of scale with the rest of the drawing.

A belt pulley has been shown in fig. 19.4 with alternative methods of dimensioning the diameters. Diameter is indicated by the symbol 'Ø' which precedes the figure. Staggering the dimensions assists in clarity, and dimension lines can be reduced in length by adopting the alternative arrangement given.

Fig. 19.5 illustrates a locating plate with various types of holes spaced at different angles. The angular dimensions are positioned so that they can be read from the bottom of the drawing. The different drillings are listed in tabular form. If the eight holes had been equally spaced on the same pitch-circle diameter, then it would have been sufficient to

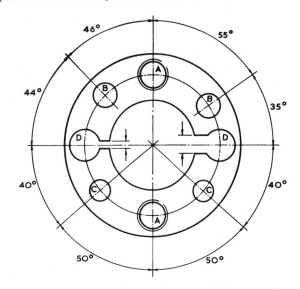

Fig. 19.5   2 holes marked 'A' tapped M15 x 1,5
2 holes marked 'B' drill Ø12
2 holes marked 'C' drill Ø10
2 holes marked 'D' drill Ø16

add a note with a leader line to one of the holes, for example '8 HOLES Ø12 EQUALLY SPACED ON A Ø125 PITCH CIRCLE', or alternatively to dimension the pitch circle and add a note to one of the holes, for example '8 HOLES Ø12 EQUISPACED'.

The dimensioning of a curved surface or profile can be arranged by two methods. Fig. 19.6 shows a template where the curves are formed from circular arcs with known radii of curvature. Centres of arcs are fixed, and the radii are quoted. This method fixes exactly every point along the curved surface and is therefore preferred. Note that the dimension lines for radii pass through the centre of arc, and that the figure is preceded by the letter 'R' in each case.

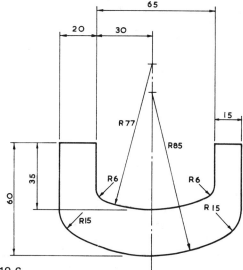

Fig. 19.6

The system of dimensioning in fig. 19.7 fixes only the points on the profile at the intersection of the rectangular co-ordinates, and is therefore less accurate. Points chosen should be as close as is possible, to keep the deviation of the curve to a minimum.

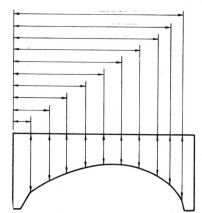

Fig. 19.7

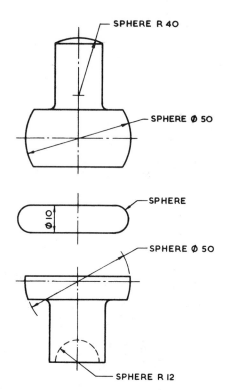

Fig. 19.8

Methods of dimensioning spherical radii and diameters are shown in fig. 19.8.

Fig. 19.9 shows various methods of dimensioning angles.

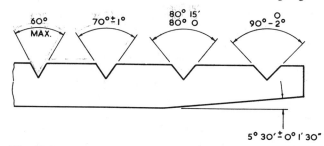

Fig. 19.9

Chapter 20

Machining and surface-roughness symbols

The quality of any finished surface has a direct connection with the function and wear of the component. Machined surface finishes vary considerably in quality, and the maximum roughness acceptable is quoted only once on the machined surface, near to the dimension fixing the location or size of the surface.

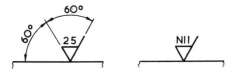

Fig. 20.1

The proportions of typical symbols are shown in fig. 20.1. The surface finish quoted on a detail drawing will, unless a note is added to the contrary, be the final finish required for the correct functioning of the part in service. The component may undergo several stages in manufacture, for example turning then grinding or chrome-plating after pressing; the roughness at the intermediate stage is normally not quoted. If only machining is required to be indicated, without reference to roughness, then the number (25 or N11 in the example above) is omitted.

In certain circumstances a surface is not required to be too smooth or too rough, and a maximum and minimum value is then quoted, as in fig. 20.2.

Fig. 20.2 Maximum and minimum values of surface roughness

Certain components are required to be machined to the same quality of finish on every surface, and, to avoid the repeated placing of the symbol on the drawing, a single general note may be added, as in fig. 20.3.

Fig. 20.3 Symbol denoting 'MACHINE ALL OVER'

Where a specific process or production method is required to be used to obtain a certain finish, then the machining symbol is slightly modified to make a reference to the fact. The horizontal line is added and the process is indicated as in fig. 20.4. Such information is normally left to the discretion of the production or methods engineer.

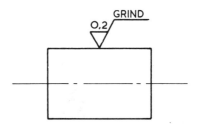

Fig. 20.4

Finished surfaces are not only obtained by machining; components with acceptable finished surfaces can be manufactured by die casting and pressing, for example. If a control is required on the quality of a surface that does not necessarily need machining, then the horizontal line on the symbol is omitted (fig. 20.5).

Fig. 20.5 Symbol where machining is not necessary

In some applications friction, for example, may be desirable, and the machining of a component already produced by another process is definitely not required. The letter 'O' is then inserted in the vee of the machining symbol, to signify the omission, as in fig. 20.6; alternatively a note 'DO NOT MACHINE' may be used.

Fig. 20.6 Symbols to indicate that machining is not required

The marks of machining are inevitable but not always undesirable; they can be used to assist in the correct functioning of parts. These marks are a succession of minute hills and valleys, and mating parts can be arranged, if necessary, with machining marks in line with one another or at other angles. Machining marks can also be used as a form of decoration on finished surfaces. If this special requirement is desired, the direction of lay must be indicated in association with the machining symbol. The end of the shaft in fig. 20.7 will be grooved like a gramophone record. The surface of the block in fig. 20.8 will have minute grooves parallel with the line of the arrow.

Fig. 20.9 shows approximate surface-roughness ranges for components manufactured by some common production methods. This information can only be approximate, since finish depends on many factors, such as the skill of the machinist, the accuracy and condition of the machine,

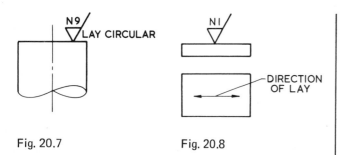

Fig. 20.7   Fig. 20.8

the speeds and feeds selected for the operation, and the quality and condition of the cutting tools.

The approximate relationship between surface roughness and the cost of producing such a finish is shown in fig. 20.10. The cost of rough machining can be considered as the zero datum on the y axis of the graph, and other processes can be compared with it. For example, a finish of 6.3 micrometres, or roughness N9, produced by grinding may well cost four times as much as rough machining. Many factors contribute towards production costs, and this information again can be only approximate.

Fig. 20.9   Approximate surface roughness heights obtainable by various common production processes.

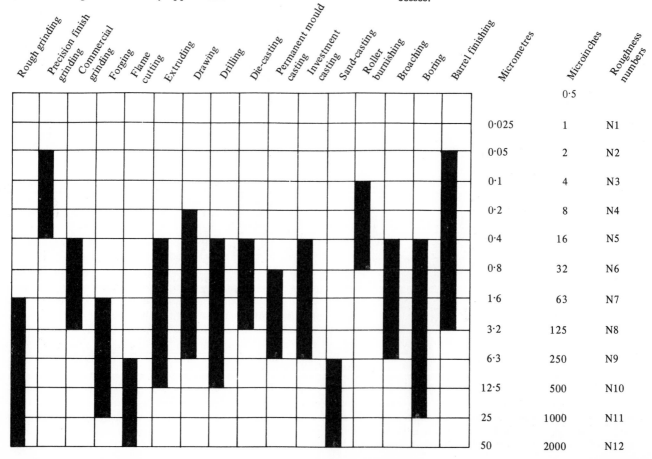

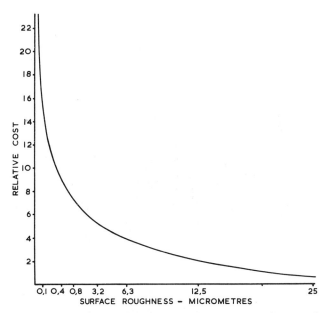

Fig. 20.10 Approximate relationship between surface roughness and cost.

Fig. 20.9 (cont'd)

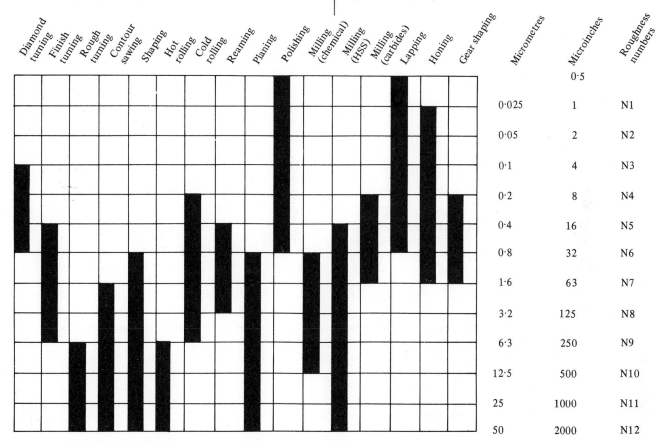

Micrometres	Microinches	Roughness numbers
	0·5	
0·025	1	N1
0·05	2	N2
0·1	4	N3
0·2	8	N4
0·4	16	N5
0·8	32	N6
1·6	63	N7
3·2	125	N8
6·3	250	N9
12·5	500	N10
25	1000	N11
50	2000	N12

Chapter 21

Screw-threads

A screw-thread is formed by cutting a continuous helical groove around a cylindrical external or internal surface. Screw-threads generally are of two distinct types: 'V' threads and square threads.

Screw-threads may be left- or right-hand, according to the direction of the helix, and a screw may have one or more threads which are cut side by side. A single-cut right-hand thread is the most common, but, where the nut is required to move an increased distance along the axis of rotation for one revolution, multiple threads can be used.

Frictional resistance to motion is less with the square thread, and this type of thread is widely used for power transmission. The 'V' thread, with its greater frictional resistance, is used for nuts and bolts and similar applications.

Thread terms

1 **Pitch** The pitch of a thread is the distance, measured along the axis of the thread, between corresponding points on adjacent threads. For a single-start thread, it will be equal to the distance moved by the nut during one revolution.

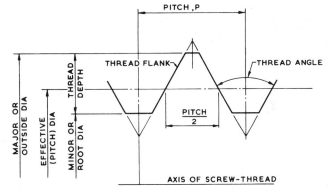

Fig. 21.1   Thread terms

2 **Lead** The lead of a thread is the distance moved by the nut in one revolution. For a single-start thread, pitch and lead have the same value, but for a two-start thread the lead will be twice the pitch, and a three-start thread will have a lead three times the pitch.
3 **Effective diameter** is the diameter where the width of the tooth is equal to the space between successive teeth. It is often known as the *pitch diameter*.
4 **Root** The root of the thread is the bottom of the groove for a male or a female thread.
5 **Crest** The crest of the thread is the most prominent part of a male or a female thread.
6 **Flank** The flank of a thread joins the crest of a thread to its root.
7 **Thread angle** The thread angle is measured between the flanks of the thread.

8 **Root diameter** The root diameter is the smallest diameter of a male thread, and is sometimes known as the *core diameter* or *minor diameter*.
9 **Outside diameter** The outside diameter is the greatest diameter of a male thread, often referred to as the *major diameter*.

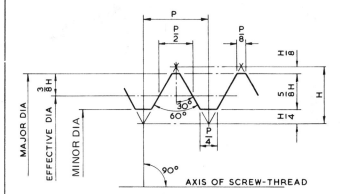

Fig. 21.2   ISO metric screw-threads
$H = 0.86603P$, $H/4 = 0.21651P$, $H/8 = 0.10825P$, $3H/8 = 0.32476P$, $5H/8 = 0.54127P$.

Fig. 21.2 shows the basic ISO metric thread form. BS 3643 defines two series of diameters with graded pitches for general use in nuts, bolts, and screwed fittings, one series with coarse and the other with fine pitches. The extract given below from Table 2 of the standard gives thread sizes from 1.6 to 24mm diameter. Note that first, second, and third choices of basic diameters are quoted, to limit the number of sizes within each range.

On a drawing, a thread will be designated by the letter M followed by the size of the nominal diameter and the pitch required, e.g. M10 x 1.5.

If a thread is dimensioned without reference to the pitch, e.g. M16, then it is assumed that the coarse-series thread is required.

Other thread forms in common use are as follows.
Acme thread (fig. 21.3). Often used as a leadscrew for lathes, this thread is adapted from the square thread form.

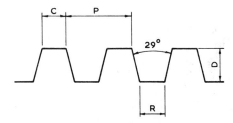

Fig. 21.3   Acme thread
$C = 0.3707P$, $R = C$, $D = P/2 + 0.01$

1	2	3	4	5	6	7	8	9	10	11	12	13	14	15	16
Basic major diameters			Coarse series with graded pitches	Pitches											
				Fine series with constant pitches											
First choice	Second choice	Third choice		6	4	3	2	1·5	1·25	1	0·75	0·5	0·35	0·25	0·2
1·6	—	—	0·35	—	—	—	—	—	—	—	—	—	—	—	0·2
—	1·8	—	0·35	—	—	—	—	—	—	—	—	—	—	—	0·2
2	—	—	0·4	—	—	—	—	—	—	—	—	—	—	0·25	—
—	2·2	—	0·45	—	—	—	—	—	—	—	—	—	—	0·25	—
2·5	—	—	0·45	—	—	—	—	—	—	—	—	—	0·35	—	—
3	—	—	0·5	—	—	—	—	—	—	—	—	—	0·35	—	—
—	3·5	—	0·6	—	—	—	—	—	—	—	—	—	0·35	—	—
4	—	—	0·7	—	—	—	—	—	—	—	—	0·5	—	—	—
—	4·5	—	0·75	—	—	—	—	—	—	—	—	0·5	—	—	—
5	—	—	0·8	—	—	—	—	—	—	—	—	0·5	—	—	—
—	—	5·5	—	—	—	—	—	—	—	—	—	0·5	—	—	—
6	—	—	1	—	—	—	—	—	—	—	0·75	—	—	—	—
—	—	7	1	—	—	—	—	—	—	—	0·75	—	—	—	—
8	—	—	1·25	—	—	—	—	—	—	1	0·75	—	—	—	—
—	—	9	1·25	—	—	—	—	—	—	1	0·75	—	—	—	—
10	—	—	1·5	—	—	—	—	—	1·25	1	0·75	—	—	—	—
—	—	11	1·5	—	—	—	—	—	—	1	0·75	—	—	—	—
12	—	—	1·75	—	—	—	—	1·5	1·25	1	—	—	—	—	—
—	14	—	2	—	—	—	—	1·5	1·25*	1	—	—	—	—	—
—	—	15	—	—	—	—	—	1·5	—	1	—	—	—	—	—
16	—	—	2	—	—	—	—	1·5	—	1	—	—	—	—	—
—	—	17	—	—	—	—	—	1·5	—	1	—	—	—	—	—
—	18	—	2·5	—	—	—	2	1·5	—	1	—	—	—	—	—
20	—	—	2·5	—	—	—	2	1·5	—	1	—	—	—	—	—
—	22	—	2·5	—	—	—	2	1·5	—	1	—	—	—	—	—
24	—	—	3	—	—	—	2	1·5	—	1	—	—	—	—	—

Note. For preference, choose the diameters given in Column 1. If these are not suitable, choose from Column 2, or finally from Column 3.

* The pitch of 1·25 mm for 14 mm diameter is to be used only for sparking plugs.

BS 3643: Part 1, 1963. Table 2—ISO metric screw threads, standard series (dimensions in millimetres)

Buttress thread (fig. 21.4). Designed for use where load is transmitted, mainly in one direction, e.g. in vice spindles.

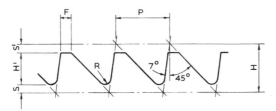

Fig. 21.4   Buttress thread
$H = 0.8906P, H' = 0.5058P, S = 0.1395P$
$S' = 0.245P, F = 0.2754P, R = 0.1205P$

Sellers or American thread (fig. 21.5). This type was the American National thread in common use before the introduction of the Unified National thread, as it is described in the USA and Canada, or the Unified screw-thread in Great Britain. The *Unified* thread form (fig. 21.6) was agreed in 1948 by a standardisation committee of the United Kingdom, the United States, and Canada, and this thread is covered by BS 1580 in three classes of fit. Note that the forms of the thread for the nut and for the bolt differ in the shape of the root and crest.

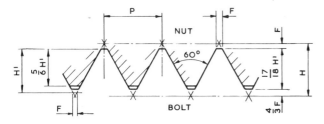

Fig. 21.5   Sellers or American thread
$H = 0.866P, H' = 0.6495P, F = 0.1083P = H/8 = H'/6$

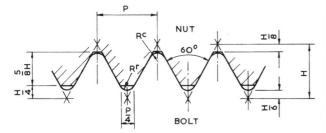

Fig. 21.6   Unified screw-thread
$H = 0.866P, R^C = 0.108P, R^r = 0.144P$

Square thread (fig. 21.7). Used to transmit force and motion since it offers less resistance to motion than 'V' thread forms. Applied widely on lathes and valve spindles.

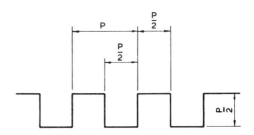

Fig. 21.7   Square thread

Whitworth thread (fig. 21.8). The general shape of the thread shown has been used in a standard BSW thread, in fine form as the BSF thread, and as a pipe thread in the BSP thread.

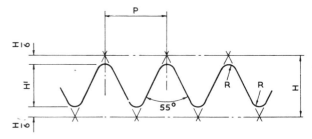

Fig. 21.8   British Standard Whitworth (BSW) thread
$H = 0.9605P, H' = 0.6403P, R = 0.1373P$

a) The *British Standard Whitworth* thread was the first standardised British screw-thread.

b) The *British Standard Fine* thread is of Whitworth section but of finer pitch. The reduction in pitch increases the core diameter; also, small adjustments of the nut can easily be made.

c) The *British Standard Pipe* threads are used internally and externally on the walls of pipes and tubes. The thread pitch is relatively fine, so that the tube thickness is not unduly weakened.

British Association thread (fig. 21.9). Generally used in sizes of less than ¼ inch on small mechanisms. This range of threads extends down to a thread size of 0.25mm and is covered by BS 93.

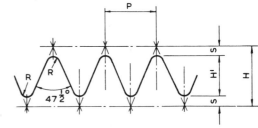

Fig. 21.9   British Association (BA) thread
$H = 1.1363P, H' = 0.6P$ (approx.), $R \cdot 0.18P,$
$S = 0.268P$

Chapter 22

Conventional representation of screw-threads

Drawing conventions are required for screw-threads, but the conventions themselves do not differentiate between different types of thread. It is therefore necessary, when dimensioning, to define precisely the thread type, size, and grade.

Fig. 22.1 shows a stud and the conventional representation of a male thread. In the end view, the projection of the major and minor diameters gives concentric circles, and, to show that the thread is male, the minor-diameter circle is drawn with a small gap. Note in the elevation that the end of the thread runs out at an angle of 30°. This small point is a modification from the previous standard.

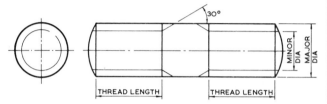

Fig. 22.1

A section through a tapped hole is shown in fig. 22.2, to indicate the conventional representation of a female thread. Note that, at the bottom of the hole, the cutting edges of the drill will leave a conical shape which is drawn with sides at 30°. The end of the threaded portion also runs out at an angle of 30°. Since the hole does not contain a male thread, the section lines are taken up to the hole left by the tapping drill.

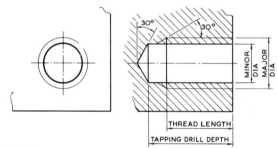

Fig. 22.2

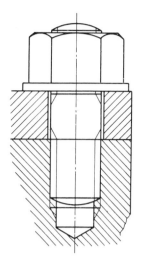

Fig. 22.3

The elevation of the female thread above shows concentric circles, and, to differentiate from the male convention, the outer circle is drawn with a small gap. In both conventional representations, the complete circle is drawn with a 0.7 line and the broken circle with a 0.3 line.

A section through a typical hole containing a stud is shown in fig. 22.3. The section lines are taken only to the side of the stud in an assembly.

Chapter 23

Nuts, bolts, screws, and washers

Nuts and bolts are drawn so often that it is necessary to use a quick easy method to obtain the approximate shape, after the designer has selected the correct size for the assembly.

The following method can be used, and will be found to cover the sizes of hexagon nuts and bolts in the Metric, Unified, and Whitworth ranges; however, for the exact size of any particular nut or bolt, it will be necessary to refer to the appropriate British Standard.

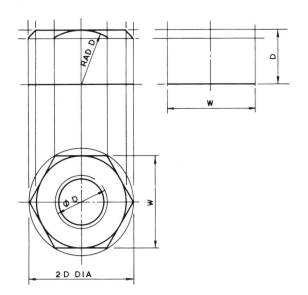

Fig. 23.1   Stage 1

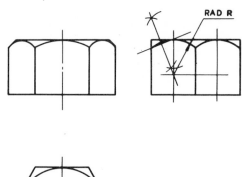

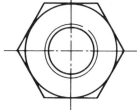

Fig. 23.2   Stage 2

Stage 1 (fig. 23.1)

1 Draw a circle, $2D$ dia, in the plan view, where D is equal to the shank diameter of the bolt and the nominal diameter of the thread.
2 Draw a hexagon inside the circle.
3 Draw a circle as shown inside the hexagon. This circle is the projection of the chamfer on the front elevation, to remove the sharp corners.
4 The nut thickness is equal to D in the front elevation. Project the four corners of the hexagon in the front elevation from the plan view, also the width W of the hexagon in the end elevation.
5 The process of chamfering will produce curves in the front and end elevations. Draw the curve shown of radius D, and project a line across each view from the bottom of the curve.
6 The female-thread convention is added to the plan.
7 Draw the projected diameter of the chamfer circle in the front elevation, and line in the chamfer as shown.

Stage 2 (fig. 23.2)

1 Note that the end elevation of the nut has sharp corners and that the corner which coincides with the centre line finishes at the bottom of the chamfer curve. The chamfer curve on the remaining four sides in both views is not part of a perfect circle, since the sides concerned all slope back from the surface of the paper; however, it is normal practice to draw the curves as circular arcs, and the construction to find the centre of the arc in the end elevation is shown.
2 Draw the centre line on the sloping face of the hexagon.
3 Bisect the chord as shown, and extend the bisector to intersect with the centre line. Radius R gives the chamfer circle on this face. Repeat to complete the end elevation.
4 This procedure would have to be repeated in the front elevation where smaller curves are required.
5 Now, a draughtsman rarely follows this procedure, but merely estimates the centre of arc. There is clearly only one arc, and, with a little practice, 'trial and error' soon becomes the application of skill and judgement. Complete the front elevation as shown.

The proportions of hexagonal nuts and bolts are shown in fig. 23.3. Note that the thicknesses of the nut, bolthead, and locknut differ, and that the quoted thicknesses of D, $0.75D$, and $0.5D$ will cover most applications. The approximate diameter of a washer is $2D + 5\text{mm}$, and its thickness $0.1D$. Washers may be plain or chamfered, as shown. Note that the projections of the curves due to the chamfers in the bolthead and locknut are the same as for the nut. The procedure is similar; only the thickness changes in each case.

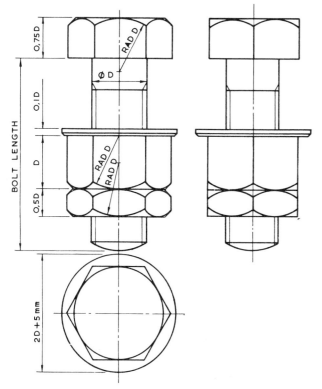

Fig. 23.3

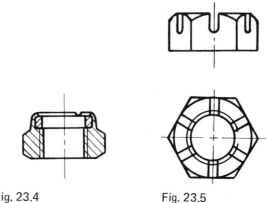

Fig. 23.4   Fig. 23.5

application, but the slots are spaced around a circular crown of reduced diameter on top of a standard hexagon nut.

Setscrews

This term is generally applied to screws where the whole shank is threaded. Apart from hexagon heads, the types shown in fig. 23.6 are the most regularly used, and are available in most thread forms.

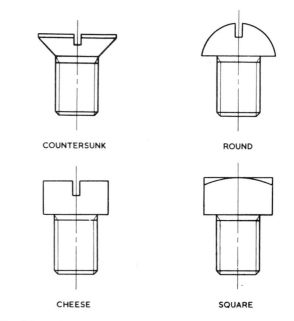

COUNTERSUNK   ROUND

CHEESE   SQUARE

Fig. 23.6   Setscrews

Locknuts

In addition to the hexagon nuts and locknuts previously shown, there are many special types of nuts which are designed and used for locking purposes. For any particular application, consideration should be given to the following circumstances.

a) *Vibration.* In cases of severe vibration, a locknut should be selected with a large frictional area.

b) *Temperature.* For some high-temperature applications, plastics and other non-metallic inserts are unsuitable.

c) *Reliability.* There is no uniformity in the effectiveness of locking of different types of locknuts. Exhaustive tests may be needed.

d) *Assembly.* Some nuts are reusable (slotted nuts) but difficult to apply where access is limited; other nuts, which depend on deformation, can be used only once. Plain hexagon locknuts are only effective when they are adequately tightened; otherwise they are free to spin off—and often do. Fig. 23.4 shows a section through a nut in which a metallic, rubber, fibre, or plastics insert is used. When the nuts are fitted, the insert material collapses against the thread, and an effective friction lock is provided. Fig. 23.5 shows a slotted nut. A split-pin passes through the slots and a hole drilled through the bolt. Castle nuts are similar in

Grub screws

Grub screws are used to prevent relative rotation or sliding between two parts which depend on a light push or sliding fit for assembly purposes.

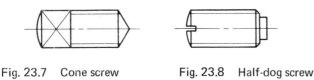

Fig. 23.7   Cone screw   Fig. 23.8   Half-dog screw

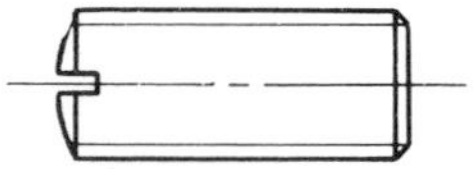

Fig. 23.9   Flat screw

a) A cone screw (fig. 23.7) can be used to bite into a circular shaft. The head is shown with a square, but it can be supplied with a slot or hexagonal recess for use with an Allen key.

b) A half-dog screw (fig. 23.8) can engage in a small slot or keyway to improve resistance to sliding or to positively prevent rotation. Optional head finishes are available.

c) Flat grub screws (fig. 23.9) are used to exert a small force on flat surfaces, without causing damage, to resist movement. These screws sometimes have a small cup-shape machined into the bottom surface, so that contact is made by the area of a ring, as distinct from the area of the circle shown in the example.

Thread-cutting screws

Barber and Colman Ltd are the manufacturers of 'Shakeproof' thread-cutting screws and washers.

'Shakeproof' thread-cutting screws made from carbon steel are subjected to a·special heat-treatment which provides a highly carburised surface with a toughened resilient core. The additional strength provided enables higher tightening torques to be used, and will often permit the use of a smaller-size thread-cutting screw than would normally be specified for a machine screw. Thread-cutting screws actually cut their own mating thread; in any thickness of material a perfect thread-fit results in greatly increased holding power, extra vibration-resistance, and a faster assembly. The hard, keen cutting edge produces a clean-cut thread, from which the screw can be removed, if desired, without damage to screw or the cut thread. The most suitable drill sizes for use with these screws are generally larger than standard tapping-drill sizes, but this apparent loss of thread engagement is more than offset by the perfect thread-fit obtained.

Both the screws shown in fig. 23.10 are interchangeable with standard machine screws. Type 1 is recommended for use in steel and non-ferrous sheet and plate, and they are manufactured with a wide shank slot and are eminently suitable for paint-clearing applications, as they completely eliminate the need for expensive pre-production tapping of painted assemblies. Type 23 screws incorporate a special

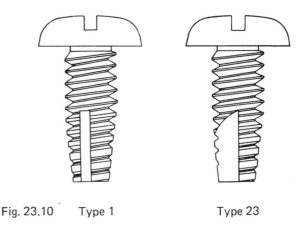

Fig. 23.10   Type 1   Type 23

wide cutting slot with an acute cutting angle for fast, easy thread-cutting action and ample swarf clearance. These screws are specially designed for application into soft metals or plastics where a standard thread form is required.

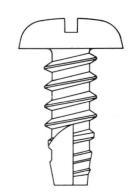

Fig. 23.11   Type 25 thread-cutting screw

The Type 25 thread-cutting screw has a specially spaced thread form which is designed for fast efficient fastening into plastics and sheet-metal applications.

Fig. 23.12 illustrates a 'Teks' self-drilling screw which, with a true drilling action, embodies three basic operations in one device. It (1) prepares its own hole, (2) either cuts or forms a mating thread, and (3) makes a complete fastening in a single operation. These screws consist of an actual drill point to which a threaded screw-fastener has been added. Several different head styles are available. During the drilling stage, Teks must be supported rigidly from the head. Some bench-mounted, automatically fed screwdrivers provide a holding means which retracts as the screw is finally driven home. Other drivers connect with the fastener only through the bit or socket. A good-fitting Phillips or Pozidriv bit will normally drive several thousand of these screws, and a hex socket, for hex-head designs, will drive even more.

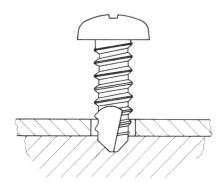

Fig. 23.12   'Teks' self-drilling screw

For long screws or applications requiring absolutely guaranteed driving stability, a special chuck is available which holds the screw with three fingers and retracts upon contacting the work surface. These screws are suitable for fastening sheet steel of 16 gauge, or thicker, within 5 seconds maximum while using a power tool operating at 2500 rev/min with 25 pounds pressure applied to the tool.

Fig. 23.13 shows alternative head styles available for thread-cutting screws.

Fig. 23.13 (a) Slotted round   Fig. 23.13 (b) Slotted pan

Fig. 23.13 (c) Slotted cheese   Fig. 23.13 (d) Slotter fillister

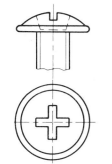 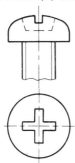

Fig. 23.13 (e) Pozidriv flange   Fig. 23.13 (f) Pozidriv pan

Fig. 23.13   Thread-cutting screws

Lockwashers and locking terminals

'Shakeproof' lockwashers have tapered twisted teeth, and work in three ways to provide an efficient method of locking threaded fasteners. This triple action, comprising strut action, line bite, and spring tension, is equally effective whether the lockwasher is used under the screw-head or the nut. In operation, these washers have sufficient resilient strength to support the load when the fastener is tightened. Resilient strength of the lockwasher is extremely important, because stress variation in a threaded fastener, caused either by vibration or by actual changes in load, is largely responsible for fatigue failure of the fastening. It has been found that, when the load variation in the fastening unit is a small percentage of the initial tightening force, the effect on the bolt is negligible; but when the load variation is a large percentage of the initial tightening force, failure incidence rises sharply. Therefore, to obtain the best results, threaded fasteners should be properly tightened, and lockwashers must be able to withstand the required tightening force and still function as a resilient lock. Fig. 23.14 shows several types of locking washer.

Lockwasher with external teeth
(a) Flat form (b) Dished type

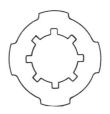 

Dished-type washer with toothed periphery

Lockwasher with   Single-coil   Tabwasher
internal teeth   washer

Fig. 23.14   Types of locking washer.

Fig. 23.15 shows a selection of locking terminals where a 'Shakeproof' washer and a soldering lug are combined into one unit, thus saving assembly time. The locking teeth anchor the terminal to the base, to prevent shifting of the terminal in handling, while the twisted teeth produce a multiple bite which penetrates an oxidised or painted surface to ensure good conductivity. All three types of locking terminal are generally made from phosphor bronze with a hot-tinned finish.

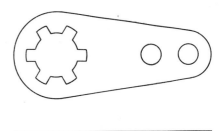

Fig. 23.15 (a) Flat type

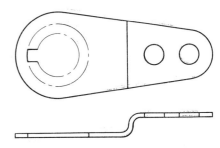

Fig. 23.15 (b) Bent type

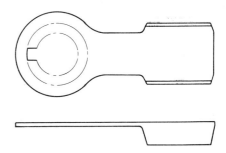

Fig. 23.15 (c) Flat-wing type

Fig. 23.15   Locking Terminals

Chapter 24

Counterbores, countersinks, and spotfaces

Counterboring

A drilling machine is used for this operation, and a typical counterboring tool is shown in fig. 24.1. The operation involves enlarging existing holes, and the depth of the enlarged hole is controlled by a stop on the drilling machine. The location of the counterbored hole is assisted

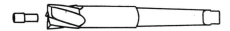

Fig. 24.1   Counterboring tool with morse taper shank (detachable—pilot pattern). Tool can also be used for spotfacing.

by a pilot at the tip of the tool which is a clearance fit in the previously drilled hole. A typical use for a counterbored hole is to provide a recess for the head of a screw, as shown in fig. 24.2, or a flat surface for an exposed nut or bolt, as in fig. 24.3. The flat surface in fig. 24.2 could also be obtained by spotfacing.

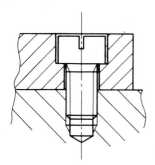

Fig. 24.2

FLAT SURFACE ON
CASTING OBTAINED
BY COUNTERBORING
OR SPOTFACING

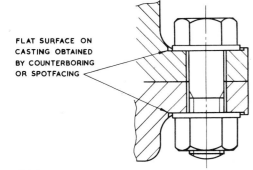

Fig. 24.3

Fig. 24.4 shows methods of dimensioning counterbores. Note that, in every case, it is necessary to specify the size of counterbore required. It is not sufficient to state 'COUNTERBORE FOR M10 RD HD SCREW', since obviously the head of the screw will fit into any counter-bore which is larger than the head.

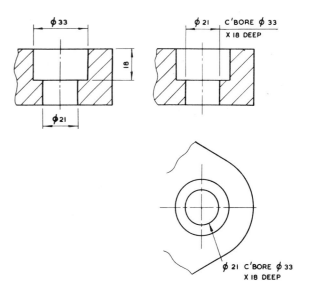

Fig. 24.4

Countersinking

Countersinking is also carried out on a drilling machine, and fig. 24.5 shows typical tools. Included angles of $60°$ and $90°$ are commonly machined, to accommodate the heads of screws and rivets to provide a flush finish (fig. 24.6).

Fig. 24.5 (a)   Taper-shank countersink (with $60°$ or $90°$ included angle of countersink)

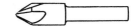

Fig. 24.5 (b)   Straight-shank machine countersink (with $60°$ or $90°$ included angle of countersink)

Spotfacing

Spotfacing is a similar operation to counterboring, but in this case the metal removed by the tool is much less. The process is regularly used on the surface of castings, to provide a flat seating for fixing bolts. A spotfacing tool is shown in fig. 24.7, where a loose cutter is used. The length of cutter controls the diameter of the spotface. As in the counterboring operation, the hole must be previously drilled, and the pilot at the tip of the spotfacing tool assists in location.

Fig. 24.8 shows the method of dimensioning. Note that, in both cases, the depth of spotface is just sufficient to remove the rough surface of the casting over the 40mm diameter area.

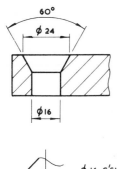

Fig. 24.6

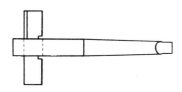

Fig. 24.7   Spotfacing tool with loose cutter

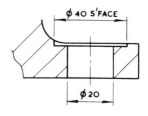

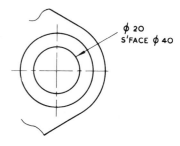

Fig. 24.8

Chapter 25

Tapers and chamfers

In fig. 25.1, the difference in magnitude between dimensions X and Y (whether diameters or widths) divided by the length between them defines a ratio known as a *taper*.

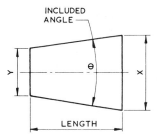

Fig. 25.1

$$\text{Taper} = \frac{X - Y}{\text{length}} \equiv 2 \tan \frac{\theta}{2}$$

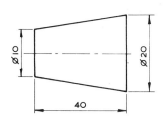

Fig. 25.2

For example, the conical taper in fig. 25.2

$$= \frac{20-10}{40} = \frac{10}{40} = 0.25$$

and may be expressed as rate of taper 0.25:1 on diameter.

The ISO recommended symbol for taper is ▭▶, and this symbol can be shown on drawings accompanying the rate of taper,

i.e. ▭▶ 0.25:1

The arrow indicates the direction of taper.

When a taper is required as a datum, it is enclosed in a box as follows:

Methods of dimensioning tapers

The size, form, and position of a tapered feature can be defined by calling for a suitable combination of the following:

1 the rate of taper, or the included angle;
2 the diameter or width at the larger end;
3 the diameter or width at the smaller end;
4 the length of the tapered feature;
5 the diameter or width at a particular cross-section, which may lie within or outside the feature concerned;
6 the locating dimension from the datum to the cross-section referred to above.

Care must be taken to ensure that no more dimensions are quoted on the drawing than are necessary. If reference dimensions are given to improve communications, then they must be shown in brackets, e.g. (1:5 taper).

Fig. 25.3 gives four examples of the methods used to specify the size, form, and position of tapered features.

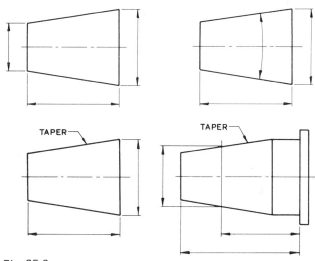

Fig. 25.3

There are three different methods of dimensioning which govern the precision of tapered features, and they can be illustrated as follows.

Basic taper method The accuracy of the rate of taper is controlled by the limits of size quoted. At any cross-section throughout the length of the feature, errors of form are contained within the tolerance limits of size.

Case 1. Product requirement (fig. 25.4)
The finished tapered surface together with any errors of form that may exist must lie within the tolerance zone.

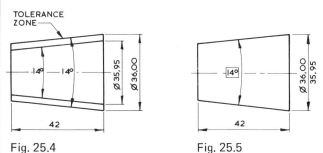

Fig. 25.4   Fig. 25.5

Case 1. Drawing instruction (fig. 25.5)
Note that the angle is boxed, which implies a true angle, i.e. an angle with no angular tolerance.

Case 2. Product requirement (fig. 25.6)
The finished tapered surface, together with any errors of form, must lie within the tolerance zone.

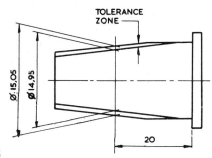

Fig. 25.6

Case 2. Drawing instruction (fig. 25.7).
Note that the tolerance zone is positioned exactly by a boxed dimension from the datum and an exact rate of taper, also boxed.

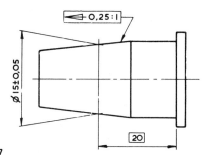

Fig. 25.7

Case 3. Product requirement (fig. 25.8)
The surface of the tapered part must lie within the tolerance zone which is fixed by an exact boxed diameter at two separate distances from the datum. An exact rate of taper is specified at these distances.

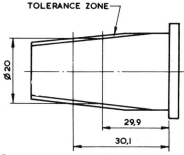

Fig. 25.8

Case 3. Drawing instruction (fig. 25.9)

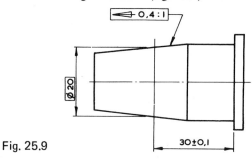

Fig. 25.9

Toleranced taper method In this method, the accuracy of the taper is controlled by the application of two separate tolerances:
1 the rate of taper tolerance,
2 the tolerance on width or diameter of the feature.
Since these two tolerances are independently applied, the extreme conditions shown in fig. 25.10 and fig. 25.11 can arise when the feature is finished at its low limit and its

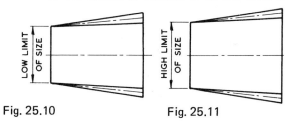

Fig. 25.10   Fig. 25.11

high limit of size. Fig. 25.12 shows the result of combining these two diagrams to give the resulting possible tolerance zone; it will be noted that the same tolerance of size does not apply at every cross-section along the feature. A typical component toleranced by this method is shown in fig. 25.13.

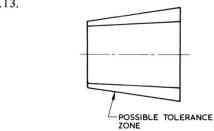

Fig. 25.12

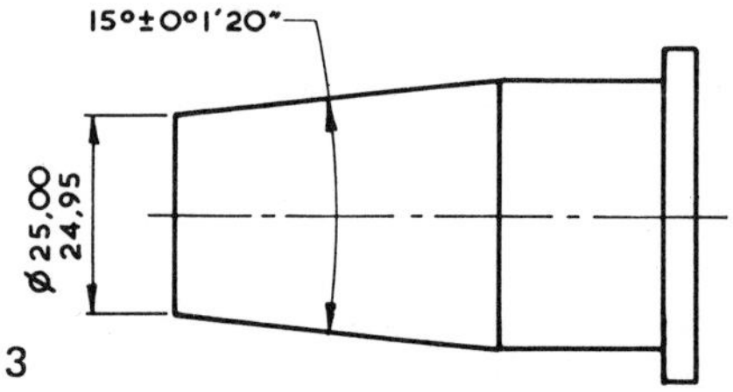

Fig. 25.13

Dimensioning two mating tapers This method is used when the fit to a mating component or gauge is necessary; the following information should then be added to the feature:
1 'TO FIT PART NO. YYY',
2 'TO FIT GAUGE (PART NO. GGG)'.
When note 2 is added to the drawing, this implies that a 'standard rubbing gauge' will give an acceptable even marking when 'blued'. The functional requirement, whether the end-wise location is important or not, will determine the method and choice of dimensioning.

An example of dimensioning two mating tapers when end-wise location is important, and adjacent dimensions are unimportant, is shown in fig. 25.14. Note that the datum diameter could be inside a female cone.

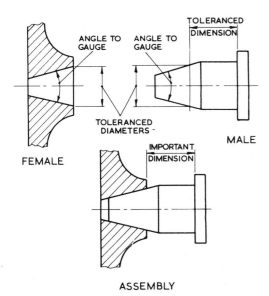

Fig. 25.14

Chamfers

Alternative methods of dimensioning internal and external chamfers are shown in fig. 25.15.

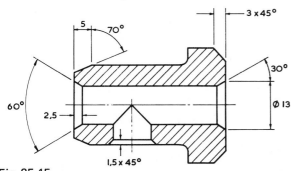

Fig. 25.15

Chapter 26

Keys and keyways

A key, fig. 26.1, is usually made from steel and is inserted between the joint of two parts to prevent relative movement; it is also inserted between a shaft and a hub in an axial direction, to prevent relative rotation. A keyway, figs 26.2, 26.3, and 26.4, is a recess in a shaft or hub to receive a key, and these recesses are commonly cut on key-seating machines or by broaching, milling, planing, shaping, and slotting. The proportions of cross-sections of keys vary with the shaft size, and reference should be made to BS 4325 for the exact dimensions. The length of the key controls the area of the cross-section subject to shear, and will need to be calculated from a knowledge of the forces being transmitted or, in the case of feather keys, the additional information of the length of axial movement required.

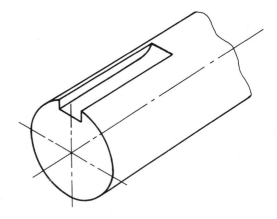

Fig. 26.2   Edge-milled keyway

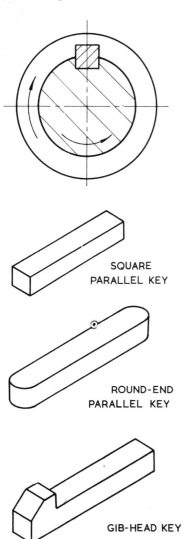

SQUARE
PARALLEL KEY

ROUND-END
PARALLEL KEY

GIB-HEAD KEY

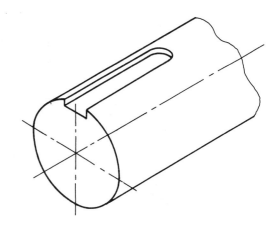

Fig. 26.3   End-milled keyway

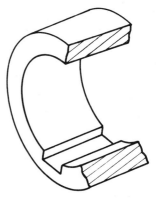

Fig. 26.1

Fig. 26.4   Keyway in hub

Sunk keys

Examples of sunk keys are shown in fig. 26.5, where the key is sunk into the shaft for half its thickness. This measurement is taken at the side of the key, and not along the centre line through the shaft axis. Fig. 26.5 shows useful proportions used for assembly drawings.

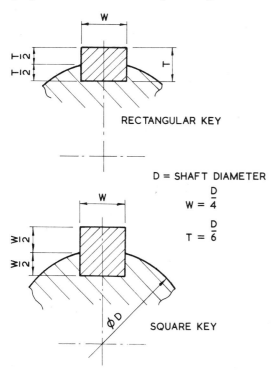

Fig. 26.5

Square and rectangular keys may be made with a taper of 1 in 100 along the length of the key; fig. 26.6 shows such an application. Note that, when dimensioning the mating hub, the dimension into the keyway is taken across the maximum bore diameter.

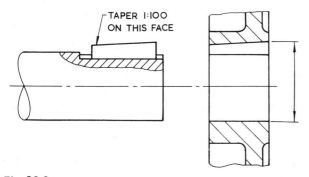

Fig. 26.6

A *gib head* may be added to a key to facilitate removal, and its proportions and position when assembled are given in fig. 26.7.

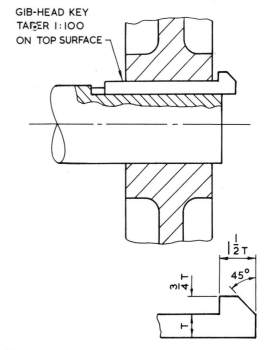

Fig. 26.7

A *feather key* is attached to either the shaft or the hub, and permits relative axial movement while at the same time enabling a twisting moment to be transmitted between shaft and hub or vice versa. Both pairs of opposite faces of the key are parallel.

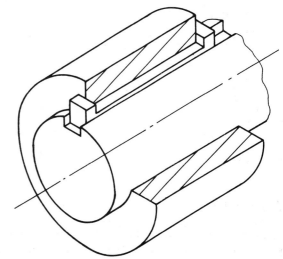

Fig. 26.8   Double-headed feather key

A *double-headed feather key* is shown in fig. 26.8 and allows a relatively large degree of sliding motion between shaft and hub. The key is inserted into the bore of the hub, and the assembly is then fed onto the shaft, thus locking the key in position.

A *peg feather key* is shown in fig. 26.9, where a peg attached to the key is located in a hole through the hub.

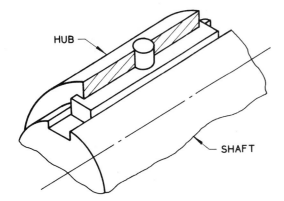

Fig. 26.9   Peg feather key

Fig. 26.10 illustrates a feather key which is screwed in position in the shaft keyway by two countersunk screws.

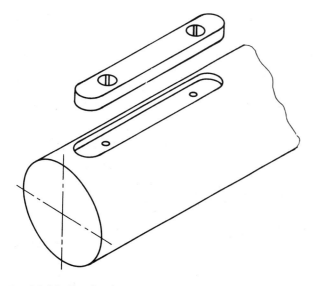

Fig. 26.10   Feather key

Woodruff keys

A Woodruff key, fig. 26.11, is a segment of a circular disc and fits into a circular recess in the shaft which is machined by a Woodruff keyway cutter. The shaft may be parallel or tapered, figs 26.12 and 26.13 showing the method of

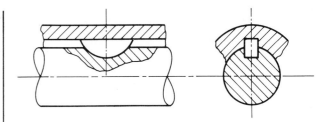

Fig. 26.11   Woodruff key

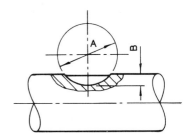

Fig. 26.12   Dimensions required for a Woodruff key in a parallel shaft

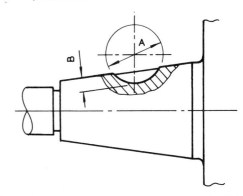

Fig. 26.13   Dimensions required for a Woodruff key in a tapered shaft

dimensioning shafts for Woodruff keys where the depth of the recess from the outside of the shaft is given, in addition to the diameter of the recess. A Woodruff key has the advantage that it will turn itself in its circular recess to accommodate any taper in the mating hub on assembly; for this reason it cannot be used as a feather key, since it would jam. Woodruff keys are commonly used in machine tools and, for example, between the flywheel and the crankshaft of a small internal-combustion engine where the drive depends largely on the fit between the shaft and the conically bored flywheel hub. The deep recess for a Woodruff key weakens the shaft, but there is little tendency for the key to turn over when in use.

Where lighter loads are transmitted and the cost of cutting a keyway is not justified, round keys and flat or hollow saddle keys as shown in fig. 26.14 can be used.

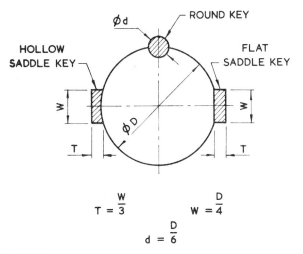

$$T = \frac{W}{3} \qquad W = \frac{D}{4}$$

$$d = \frac{D}{6}$$

Fig. 26.14

Saddle keys are essentially for light duty only, overloading tending to make them rock and work loose on the shaft. Both flat and hollow saddle keys may have a taper of 1 in 100 on the face in contact with the hub. The round key may either be tapered or, on assembly, the end of the shaft and hub may be tapped after drilling and a special threaded key be screwed in to secure the components.

Dimensioning keyways (parallel keys)

The method of dimensioning a parallel shaft is shown in fig. 26.15, and a parallel hub in fig. 26.16. Note that in each case it is essential to show the dimension to the bottom of the keyway measured across the diameter of the shaft and the bore of the hub. This practice cannot be used where either the shaft or hub is tapered, and fig. 26.17 shows the method of dimensioning a keyway for a square or rectangular parallel key in a tapered shaft, where the keyway depth is shown from the outside edge of the shaft and is measured vertically into the bottom of the slot. Fig. 26.18 shows a tapered hub with a parallel keyway where the dimension to

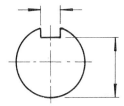

Fig. 26.15   Keyway in parallel shaft

the bottom of the slot is taken across the major diameter. A parallel hub utilising a tapered key is also dimensioned across the major diameter, as indicated in fig. 26.19.

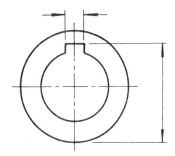

Fig. 26.16   Keyway in parallel hub

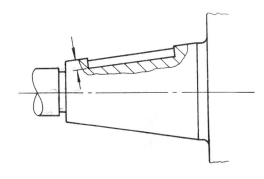

Fig. 26.17   Keyway for square or rectangular parallel key in tapered shaft

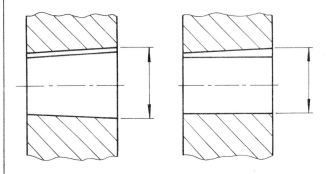

Fig. 26.18   Tapered hub with   Fig. 26.19   Parallel hub with
parallel keyway   tapered keyway

Chapter 27

Springs

Mechanical springs may be defined as elastic bodies whose primary function is to deform under a load and return to their original shape when the load is removed. In practice, the vast majority of springs are made of metal, and of these the greatest proportion are of plain-carbon steel.

Plain-carbon steels

These steels have a carbon-content ranging from about 0.5% to 1.1%. In general it may be taken that, the higher the carbon-content, the better the spring properties that may be obtained.

In the manufacture of flat springs and the heavier coil springs, it is usual to form the spring from annealed material and subsequently to heat-treat it. However, it is sometimes possible to manufacture flat springs from material which is already in the hardened and tempered condition, and this latter technique may give a lower production cost than the former.

For light coil springs, the material loosely known as piano wire is used; this is a spring wire which obtains its physical properties from cold-working, and not from heat-treatment. Springs made from this wire require only a low-temperature stress-relieving treatment after manufacture. Occasionally wire known as 'oil-tempered' wire is used—this is a wire which is hardened and tempered in the coil form, and again requires only a low-temperature stress relief after forming.

Plain-carbon steel springs are satisfactory in operation up to a temperature of about 180°C. Above this temperature they are very liable to take a permanent set, and alternative materials should be used.

Alloy steels

Alloy steels are essentially plain-carbon steels to which small percentages of alloying elements such as chromium and vanadium have been added. The effect of these additional elements is to modify considerably the steels' properties and to make them more suitable for specific applications than are the plain-carbon steels. The two widely used alloy steels are
a) chrome-vanadium steel—this steel has less tendency to set than the plain-carbon steels;
b) silicon-manganese steel—a cheaper and rather more easily available material than chrome-vanadium steel, though the physical properties of both steels are almost equivalent.

Stainless steels

Where high resistance to corrosion is required, one of the stainless steels should be specified. These fall into two categories.
a) *Martensitic*. These steels are mainly used for flat springs with sharp bends. They are formed in the soft condition and then heat-treated.

b) *Austenitic*. These steels are cold-worked for the manufacture of coil springs and flat springs, with no severe forming operations.

Both materials are used in service up to about 235°C.

High-nickel alloys

Alloys of this type find their greatest applications in high-temperature operation. The two most widely used alloys are
a) Inconel—a nickel-chrome-iron alloy for use up to about 320°C;
b) Nimonic 90—a nickel-chrome-cobalt alloy for service up to about 400°C, or at reduced stresses to about 590°C.

Both of these materials are highly resistant to corrosion

Copper-base alloys

With their high copper-content, these materials have good electrical conductivity and resistance to corrosion. These properties make them very suitable for such purposes as switch-blades in electrical equipment.
a) *Brass*—an alloy containing zinc, roughly 30%, and the remainder copper. A cold-worked material obtainable in both wire and strip form, and which is suitable only for lightly stressed springs.
b) *Phosphor bronze*—the most widely used copper-base spring material, with applications the same as those of brass, though considerably higher stresses may be used.
c) *Beryllium copper*—this alloy responds to a precipitation heat-treatment, and is suitable for springs which contain sharp bends. Working stresses may be higher than those used for phosphor bronze and brass.

Compression springs

Fig. 27.1 shows two alternative types of compression springs for which drawing conventions are used. Note that the convention in each case is to draw the first and last 1½ turns of the spring and to then link the space in between with a thin-chain line. The diagrammatic representation shows the coils of the springs drawn as single lines.

A schematic drawing of a helical spring is shown in fig. 27.2. This type of illustration can be used as a working drawing in order to save draughting time, with the appropriate dimensions and details added.

Fig. 27.3 shows four of the most popular end formations used on compression springs. When possible, grinding should be avoided, as it considerably increases spring costs.

Fig. 27.4 shows a selection of compression springs, including valve springs for diesel engines and injection pumps.

Conical compression springs with ground ends

Cylindrical compression spring with ground ends

(a) Closed ends, ground

Drawing convention

Drawing convention

(b) Closed ends

(c) Open ends, ground

Diagrammatic representation

Diagrammatic representation

(d) Open ends

Fig. 27.3

Fig. 27.1

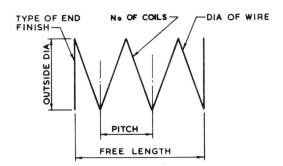

Fig. 27.2   Schematic drawing of helical spring

Fig. 27.4

Flat springs

Fig. 27.5 shows a selection of flat springs, circlips, and spring pressings. It will be apparent from the selection that it would be difficult, if not impossible, to devise a drawing standard to cover this type of spring, and at present none exists.

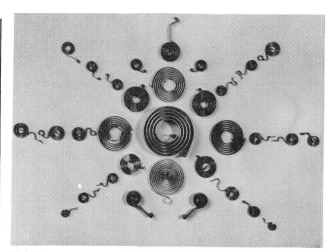

Fig. 27.6

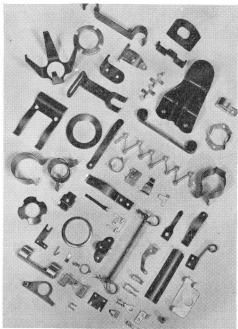

Fig. 27.5

Flat springs are usually made from high-carbon steel in the annealed condition, and are subsequently heat-treated; but the simpler types without bends can frequently be made more economically from material pre-hardened and tempered to the finished hardness required. Stainless steels are used for springs where considerable forming has to be done. For improved corrosion-resistance, 18/8 stainless steel is recommended; but, since its spring temper is obtained only by cold-rolling, severe bends are impossible. Similar considerations apply to phosphor bronze, titanium, and brass, which are hardened by cold-rolling. Beryllium copper, being thermally hardenable, is a useful material as it can be readily formed in the solution-annealed state.

Fig. 27.6 shows a selection of flat spiral springs, frequently used for brush mechanisms, and also for clocks and motors. The spring consists of a strip of steel spirally wound and capable of storing energy in the form of torque.

Torsion springs

Various forms of single and double torsion springs are illustrated in fig. 27.7.

Fig. 27.8 gives a schematic diagram for a torsion spring. This type of drawing, adequately dimensioned, can be used for detailing.

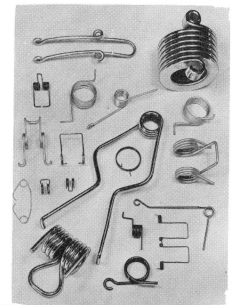

Fig. 27.7

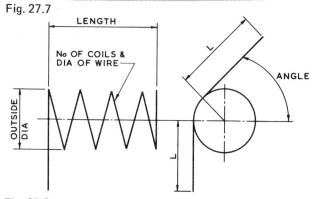

Fig. 27.8

The drawing conventions for a torsion spring are shown in fig. 27.9. Although torsion springs are available in many different forms, this is the only type to be represented in engineering-drawing standards. Torsion springs may be wound from square-, rectangular-, or round-section bar. They are used to exert a pressure in a circular arc, for example in a spring hinge and in door locks. The ends of the wire in the spring may be straight, curved, or kinked.

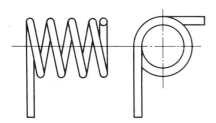

Cylindrical torsion spring

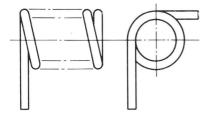

Spring convention for use on drawings

Diagrammatic representation

Fig. 27.9   Conventional representation of torsion springs

Leaf springs

The two standards applicable to leaf springs are shown in fig. 27.10. These springs are essentially strips of flat metal formed in an elliptical arc and suitably tempered. They absorb and release energy, and are commonly found applied to suspension systems.

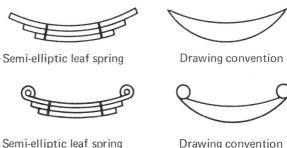

Semi-elliptic leaf spring   Drawing convention

Semi-elliptic leaf spring   Drawing convention
with fixing eyes

Fig. 27.10

Helical extension springs

A helical extension spring is a spring which offers resistance to extension. Almost invariably they are made from circular-section wire, and a typical selection is illustrated in fig. 27.11.

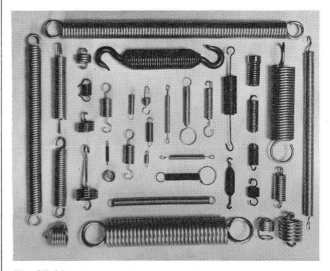

Fig. 27.11

The conventional and diagrammatic representations of tension springs are shown in fig. 27.12, and a schematic drawing for detailing is shown in fig. 27.13.

Coils of extension springs differ from those of compression springs in so far as they are wound so close together that a force is required to pull them apart. A variety of end loops is available for tension springs, and some of the types are illustrated in fig. 27.14.

A common way of reducing the stress in the end loop is to reduce the moment acting by making the end loop smaller than the body of the spring, as shown in fig. 27.15.

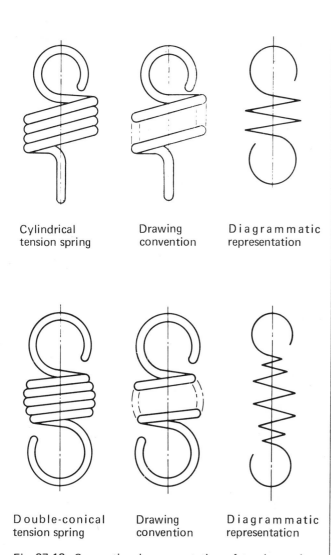

Cylindrical
tension spring

Drawing
convention

Diagrammatic
representation

Double-conical
tension spring

Drawing
convention

Diagrammatic
representation

Fig. 27.12 Conventional representation of tension springs

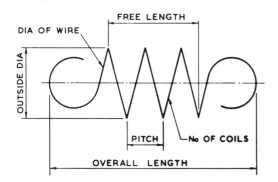

Fig. 27.13 Schematic drawing of tension spring

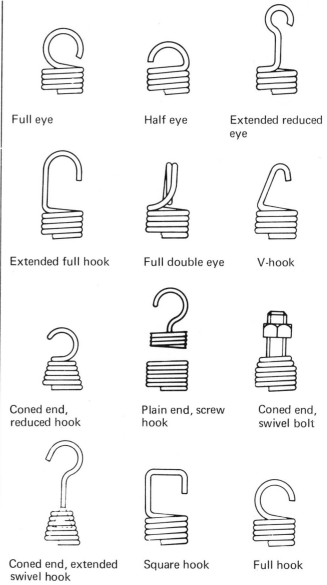

Full eye

Half eye

Extended reduced
eye

Extended full hook

Full double eye

V-hook

Coned end,
reduced hook

Plain end, screw
hook

Coned end,
swivel bolt

Coned end, extended
swivel hook

Square hook

Full hook

Fig. 27.14 Types of end loops

Fig. 27.15

Spring specifications

A frequent cause of confusion between the spring supplier and the spring user is lack of precision in specifying the spring. This often results in high costs due to the manufacturer taking considerable trouble to meet dimensions which are not necessary for the correct functioning of the spring.

It is recommended that, while all relevant data regarding the design should be entered on the spring detail drawing, some indication should be given as to which of the particular design points must be adhered to for the satisfactory operation of the component; this is necessary to allow for variations in wire diameter and elastic modulus of the material. For example, a compression spring of a given diameter may, if the number of coils is specified, have a spring rate which varies between ± 15% of the calculated value. In view of this, it is desirable to leave at least one variable for adjustment by the manufacturer, and the common convenient variable for general use is the number of coils.

A method of spring specification which has worked well in practice is to insert a table of design data, such as that shown below, on the drawing. All design data is entered, and the items needed for the correct functioning of the spring are marked with an asterisk. With this method the manufacturer is permitted to vary any unmarked items, as only the asterisked data is checked by the spring user's inspector. The following are specifications typical for compression, tension, and torsion springs.

Compression spring

Total turns	7
Active turns	5
Wire diameter	1 mm
*Free length	12.7 ± 0.4 mm
*Solid length	7 mm max.
*Outside coil diameter	7.6 mm max.
*Inside coil diameter	5 mm
Rate	7850 N/m
*Load at 9 mm	31 ± 4.5 N
Solid stress	881 N/mm^2
*Ends	Closed and ground
Wound	Right-hand or left-hand
*Material	S202
*Protective treatment	Cadmium-plate

Tension spring

Mean diameter	11.5 mm
*O.D. max.	13.5 mm
*Free length	54 ± 0.5 mm
Total coils on barrel	16½
Wire diameter	1.62 mm
*Loops	Full eye, in line with each other and central with barrel of spring
Initial tension	None
Rate	2697 N/m
*Load	53 ± 4.5 N
*At loaded length	73 mm
Stress at 53 N	438 N/mm^2
Wound	Right-hand or left-hand
*Material	BS 1408B
*Protective treatment	Lanolin

Torsion spring

Total turns on barrel	4
Wire diameter	2.6 mm
*Wound	Left-hand close coils
Mean diameter	12.7 mm
*To work on	9.5 mm dia bar
*Length of legs	28 mm
*Load applied at	25.4 mm from centre of spring
*Load	41 ± 2 N
*Deflection	20°
Stress at 4I N	595 N/mm^2
*Both legs straight and tangential to barrel	
*Material	BS 1408D
*Protective treatment	Grease

Wire forms

Many components are assembled by the use of wire forms which are manufactured from piano-type wire. Fig. 27.16 shows a selection, though the number of variations is limitless.

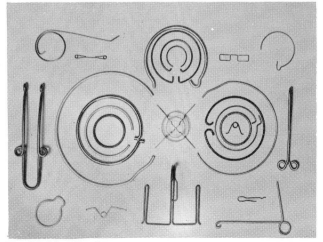

Fig. 27.16 Wire forms

Corrosion prevention

Springs operating under severe corrosive conditions are freqently made from phosphor bronze and stainless steel, and occasionally from nickel alloys and titanium alloys. For less severe conditions, cadmium- or zinc-plating is recommended; alternatively, there are other electroplatings available, such as copper, nickel, and tin. Phosphate coatings are freqently specified. Organic coatings, originally confined to stove enamels, now include many plastics materials such as nylon and polythene, as well as many types of epoxy resins.

Fatigue conditions

Many springs, such as valve springs, oscillate rapidly through a considerable range of stress, and are consequently operating under conditions requiring a long fatigue life. The suitability of a spring under such conditions must be considered at the drawing-board stage, and a satisfactory design and material must be specified. Special treatments such as shot-peening or liquid-honing may be desirable. In the process of shot-peening, the spring is subjected to bombardment by small particles of steel shot; this has the effect of work-hardening the surface. Liquid honing consists of placing the spring in a jet of fine abrasive suspended in water. This has a similar effect to shot-peening, and the additional advantage that the abrasive stream removes or reduces stress raisers on the spring surface.

Chapter 28

Welding and welding symbols

In general, welding may be described as a process of uniting two pieces of metal or alloy by raising the temperature of the surfaces to be joined so that they become plastic or molten. This may be done with or without the application of pressure and with or without the use of added metal. This definition excludes the more recently developed method of *cold-welding,* in which pressure alone is used. Cold-welding, however, has a limited application, and is used principally for aluminium and its alloys, and not for steel.

There are numerous methods of welding, but they can be grouped broadly into two categories. *Forge welding* is the term covering a group of welding processes in which the parts to be joined are heated to a plastic condition in a forge or other furnace, and are welded together by applying pressure or impact, e.g. by rolling, pressing, or hammering. *Fusion welding* is the process where the surfaces to be joined are melted with or without the addition of filler metal. The term is generally reserved for those processes in which welding is achieved by fusion alone, without pressure.

Forge welding will be dealt with first. *Pressure welding* is the welding of metal by means of mechanical pressure whilst the surfaces to be joined are maintained in a plastic state. The heating for this process is usually provided by the process of *resistance welding,* where the pieces of metal to be joined are pressed together and a heavy current is passed through them.

Projection welding is a resistance-welding process in which fusion is produced by the heat obtained from the resistance to flow of electric current through the work parts, which are held together under pressure by the electrodes providing the current. The resulting welds are localised at predetermined points by the design of the parts to be welded. The localisation is usually accomplished by projections or intersections.

Spot welding is a resistance-welding process of joining two or more overlapping parts by local fusion of a small area or 'spot'. Two copper-alloy electrodes contact either side of the overlapped sheets, under known loads produced by springs or air pressure. *Stitch welding* is spot welding in which successive welds overlap. *Seam welding* is a resistance-welding process in which the electrodes are discs. Current is switched on and off regularly as the rims of the discs roll over the work, with the result that a series of spot welds is produced at such points. If a gas-tight weld is required, the disc speed and time cycle are adjusted to obtain a series of overlapping welds.

Flash-butt welding is a resistance-welding process which may be applied to rod, bar, tube, strip, or sheet to produce a butt joint. After the current is turned on, the two parts are brought together at a predetermined rate so that discontinuous arcing occurs between the two parts to be joined. This arcing produces a violent expulsion of small particles of metal (flashing), and a positive pressure in the weld area will exclude air and minimise oxidation. When sufficient heat has been developed by flashing, the parts are brought together under heavy pressure so that all fused and oxidised material is extruded from the weld.

Fusion-welding processes can now be dealt with. The heat for fusion welding is provided by either gas or electricity. *Gas welding* is a process in which heat for welding is obtained from a gas or gases burning at a sufficiently high temperature produced by an admixture of oxygen. Examples of the gases used are acetylene (oxy-acetylene welding), hydrogen (oxy-hydrogen welding), and propane (oxy-propane welding). In *air-acetylene welding,* the oxygen is derived from the atmosphere by induction.

Electrical fusion welding is usually done by the process of 'arc welding'. *Metal-arc welding* is welding with a metal electrode, the melting of which provides the filler metal. *Carbon-arc welding* is a process of arc welding with a carbon electrode (or electrodes), in which filler metal and sometimes flux may be used. *Submerged-arc welding* is a method in which a bare copper-plated steel electrode is used. The arc is entirely submerged under a separate loose flux powder which is continually fed into and over the groove which is machined where the edges to be welded are placed together. Some of the flux powder reacts with the molten metal: part fuses and forms a refining slag which solidifies on top of the weld deposit; the remainder of the powder covers the weld and slag, shielding them from atmospheric contamination and retarding the rate of cooling.

Argon-arc welding is a process where an arc is struck between an electrode (usually tungsten) and the work in an inert atmosphere provided by directing argon into the weld area through a sheath surrounding the electrode. *Heliarc welding* uses helium to provide the inert atmosphere, but this process is not used in the United Kingdom, because of the non-availability of helium. Several proprietary names are used for welding processes of this type, e.g. *Sigma (shielded inert-gas metal-arc) welding* uses a consumable electrode in an argon atmosphere. *Atomic-hydrogen arc welding* is a process where an alternating-current arc is maintained between tungsten electrodes, and each electrode is surrounded by an annular stream of hydrogen. In passing through the arc, the molecular hydrogen is dissociated into its atomic state. The recombination of the hydrogen atoms results in a very great liberation of heat which is used for fusing together the metals to be joined. *Stud welding* is a process in which an arc is struck between the bottom of a stud and the base metal. When a pool of molten metal has formed, the arc is extinguished and the stud is driven into the pool to form a weld.

The application of welding symbols to working drawings

The notes following are meant as a guide to the method of applying the more commonly used welding symbols relating to the simpler types of welded joints to engineering drawings. BS499: part 2: 1980 gives the complete scheme. The standard is not intended to apply to complex joints involving multiple welds since it may be easier to detail such joints and welds on a seprate drawing.

Each type of weld is characterised by a symbol given in Table 28.1. Note that the symbol is representative of the shape of the weld, or the edge preparation, but does not indicate any particular welding process and does not specify either the number of runs to be deposited or whether or not a root gap or backing material is to be used. These details would be provided on a welding procedure schedule for the particular job.

It may be necessary to specify the shape of the weld surface on the drawing as flat, convex or concave and a supplementary symbol, shown in Table 28.2, is then added to the elementary symbol. An example of each type of weld surface application is given in Table 28.3.

A joint may also be made with one type of weld on a particular surface and another type of weld on the back and in this case elementary symbols representing each type of weld used are added together. The last example in Table 28.3 shows a single-V butt weld with a backing run where both surfaces are required to have a flat finish.

A welding symbol is applied to a drawing by using a reference line and an arrow line as shown in fig. 28.1. The reference line should be drawn parallel to the bottom edge of the drawing sheet and the arrow line forms an angle with the reference line. The side of the joint nearer the arrow head is known as the 'arrow side' and the remote side as the 'other side'.

The welding symbol should be positioned on the reference line as indicated in Table 28.4.

Sketch (a) shows the symbol for a single-V butt weld below the reference line because the external surface of the weld is on the arrow side of the joint.

Sketch (b) shows the same symbol above the reference line because the external surface of the weld is on the other side of the joint.

Sketch (c) shows the symbol applied to a double-V butt weld.

Sketch (d) shows fillet welds on a cruciform joint where the top weld is on the arrow side and the bottom weld is on the other side.

The positioning of the symbol is the same for drawings in first or third angle projection.

Form of weld	Illustration	BS symbol
Butt weld between flanged plates (the flanges being melted down completely)		⏝⏝
Square butt weld		‖
Single-V butt weld		V
Single-bevel butt weld		V
Single-V butt weld with broad root face		Y
Single-bevel butt weld with broad root face		Y
Single-U butt weld		Y
Single-J butt weld		P
Backing or sealing run		⏝
Fillet weld		◺
Plug weld (circular or elongated hole, completely filled)		⊓
Spot weld (resistance or arc welding) or projection weld (a) Resistance (b) Arc		○
Seam weld		⊖

Table 28.1   Elementary weld symbols

Shape of weld surface	BS symbol
flat (usually finished flush)	——
convex	⌒
concave	⌣

Table 28.2   Supplementary symbols

Form of weld	Illustration	BS symbol
Flat (flush) single-V butt weld		▽
Convex double-V butt weld		⊗
Concave fillet weld		⊿
Flat (flush) single-V butt weld with flat (flush) backing run		⊻

Table 28.3   Some examples of the application of supplementary symbols

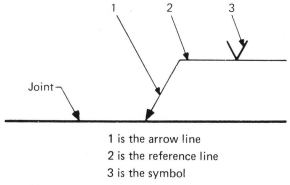

1 is the arrow line
2 is the reference line
3 is the symbol

Fig. 28.1

Additional symbols can be added to the reference line as shown in fig. 28.2. Welding can be done in the factory or on site when plant is erected. A site weld is indicated by a flag. A continuous weld all round a joint is shown by a circle at the intersection of the arrow and the reference line. Note that if a continuous weld is to be undertaken at site then both symbols should be added to the drawing.

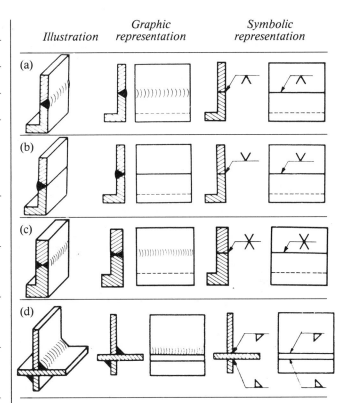

Illustration	Graphic representation	Symbolic representation
(a)		
(b)		
(c)		
(d)		

Table 28.4   Significance of the arrow and the position of the weld symbol

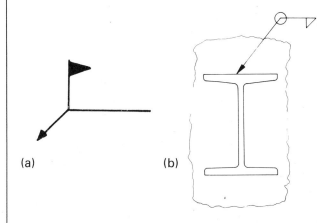

Fig. 28.2   Indication of (a) site welds and (b) continuous welds

If it is necessary to indicate on a drawing that non-destructive testing of a joint is to be undertaken then the following symbol should be added near to the end of the reference line.

A welding procedure sheet will usually give details of the actual welding process to be used on a particular joint. BS499: part 2: 1980 lists all of the current methods and if it is necessary to indicate the process as part of the symbolic representation then the appropriate number from the list in the standard is added to the drawing within a 'fork' at the end of the reference line, as shown below.

Dimensions of welds

The dimensions of a weld may be added to a drawing in the following manner.

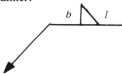

Dimensions relating to the cross section of the weld are symbolised by b and are given on the left-hand side of the symbol. The cross-sectional dimension to be indicated for a fillet weld is the leg length. If the design throat thickness is to be indicated then the leg-length dimension is prefixed with the letter b and the design throat thickness with the letter a.

Longitudinal dimensions are symbolised by l and are given on the right-hand side of the symbol. If the weld is not continuous then distances between adjacent weld elements are indicated in parentheses. Unless dimensional indication is added to the contrary, a fillet weld is assumed to be continuous along the entire length of the weld.

Leg-length dimensions of fillet welds of 3,4,5,6,8,10,12,16, 18,20,22 and 25 millimetres are the preferred sizes.

Applications of dimensions to different types of fillet welds are given in Table 28.5 in order to indicate the scope of the British Standard, which should be consulted to fully appreciate this topic. Table 28.5(a) shows dimensions applied to continuous fillet welds, (b) shows dimensions applied to intermittent fillet welds and (c) shows dimensions applied to staggered intermittent fillet welds.

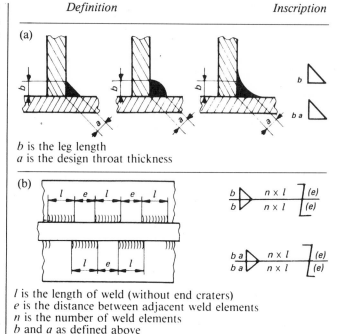

Table 28.5   The dimensioning of welds

Chapter 29

Limits and fits

To ensure that an assembly will function correctly, its component parts must fit together in a predictable manner. Now, in practice, no component can be manufactured to an exact size, and one of the problems facing the designer is to decide the upper and lower limits of size which are acceptable for each of the dimensions used to define shape and form and which will ensure satisfactory operation in service. For example, a dimension of 10±0.02 means that a part will be acceptable if manufactured anywhere between the limits of size of 9.98 mm and 10.02 mm. The present system of manufacture of interchangeable parts was brought about by the advent of and the needs of mass production, and has the following advantages.

1 Instead of 'fitting' components together, which requires some adjustment of size and a high degree of skill, they can be 'assembled'.

2 An assembly can be serviced by replacing defective parts by components manufactured to within the same range of dimensions.

3 Parts can be produced in large quantities, in some cases with less demand on the skill of the operator. Invariably this requires the use of special-purpose machines, tools, jigs, fixtures, and gauges; but the final cost of each component will be far less than if made separately by a skilled craftsman.

It should be noted, however, that full interchangeability is not always necessary in practice, neither is it always feasible, especially when the dimensions are required to be controlled very closely in size. Many units used in the construction of motor vehicles are assembled after an elaborate inspection process has sorted the components into different groups according to size. Suppose, for example, that it was required to maintain the clearance between a piston and a cylinder to within 0.012 mm. To maintain full interchangeability would require both the piston and the cylinder bores to be finished to a tolerance of 0.006 mm, which would be difficult to maintain and also uneconomic to produce. In practice it is possible to manufacture both bores and pistons to within a tolerance of 0.06 mm and then divide them into groups for assembly; this involves the gauging of each component.

A designer should ensure that the drawing conveys clear instructions regarding the upper and lower limits of size for each dimension, and figs 29.1 to 29.4 show typical methods in common use.

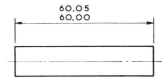

Fig. 29.1

The method shown in fig. 29.1 is perhaps the clearest method of expressing limits of size on a drawing, since the upper and lower limits are quoted, and the machine operator is not involved in mental arithmetic. The dimensions are quoted in logical form, with the upper limit above the lower limit and both to the same number of decimal places.

As an alternative to the method above, the basic size may be quoted and the tolerance limits added as in fig. 29.2. It is not necessary to express the nominal dimension to the same number of decimal places as the limits.

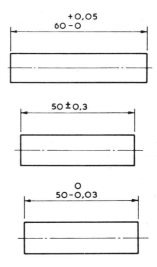

Fig. 29.2

Fits can be taken directly from those tabulated in BS 4500, 'ISO limits and fits', and, in order to indicate the grade of fit, the following alternative methods of dimensioning a hole may be used:

$$90\,\text{H7} \begin{pmatrix} 90.035 \\ 90.000 \end{pmatrix} \qquad \text{(first choice)}$$

or $\qquad 90\,\text{H7} \qquad$ or $\qquad 90\,\text{H7} \begin{pmatrix} +0.035 \\ 0 \end{pmatrix}$

Similarly, a shaft may be dimensioned as follows:

$$90\,\text{g6} \begin{pmatrix} 89.988 \\ 89.966 \end{pmatrix} \qquad \text{(first choice)}$$

or $\qquad 90\,\text{g6} \qquad$ or $\qquad 90\,\text{g6} \begin{pmatrix} -0.012 \\ -0.034 \end{pmatrix}$

In cases where a large amount of repetition is involved, information can be given in tabulated form, and a typical component drawing is shown in fig. 29.3.

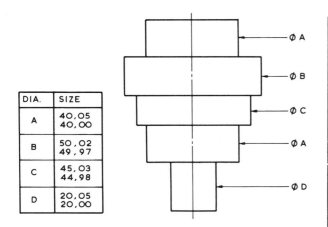

DIA.	SIZE
A	40,05 40,00
B	50,02 49,97
C	45,03 44,98
D	20,05 20,00

Fig. 29.3

In many cases, tolerances need be only of a general nature, and cover a wide range of dimensions. A box with a standard note is added to the drawing, and the typical examples in fig. 29.4 are self explanatory.

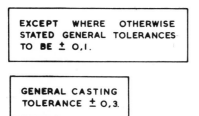

Fig. 29.4

Engineering fits between two mating parts can be divided into three types:

1 a *clearance fit* (fig. 29.5), in which the shaft is always smaller than the hole into which it fits;

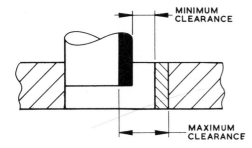

Fig. 29.5   Clearance fits—allowance always positive

2 an *interference fit* (fig. 29.6), in which the shaft is always bigger than the hole into which it fits;

3 a *transition fit* (fig. 29.7), in which the shaft may be either bigger or smaller than the hole into which it

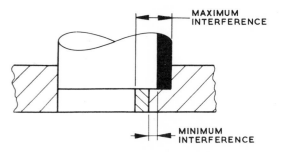

Fig. 29.6   Interference fits—allowance always negative

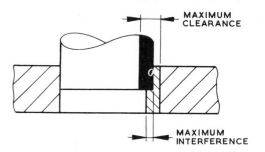

Fig. 29.7   Transition fit—allowance may be positive or negative

fits—it will therefore be possible to get interference or clearance fits in one group of assemblies.

It will be appreciated that, as the degree of accuracy required for each dimension increases, the cost of production to maintain this accuracy increases at a sharper rate.

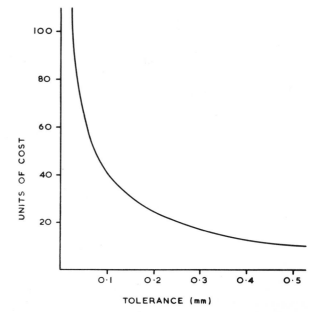

Fig. 29.8   Approximate relationship between production cost and manufacturing tolerance

Fig. 29.8 shows the approximate relationship between cost and tolerance. For all applications, the manufacturing tolerance should be the largest possible which permits satisfactory operation.

Elements of interchangeable systems (fig. 29.9)

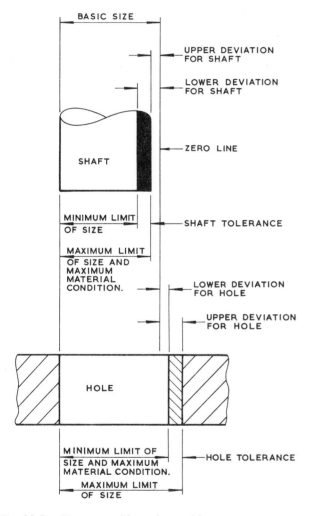

Fig. 29.9   Elements of interchangeable systems

Nominal size is the size by which a component is referred to as a matter of convenience, i.e. 25mm, 50mm, 60mm thread.

Actual size is the measured size.

Basic size is the size in relation to which all limits of size are fixed, and will be the same for both the male and female parts of the fit.

Limits of size. These are the maximum and minimum permissible sizes acceptable for a specific dimension.

Tolerance. This is the total permissible variation in the size of a dimension, and is the difference between the upper and lower acceptable dimensions.

Allowance concerns mating parts, and is the difference between the high limit of size of the shaft and the low limit of size of its mating hole. An allowance may be positive or negative.

Grade. This is an indication of the tolerance magnitude: the lower the grade, the finer will be the tolerance.

Deviation. This is the difference between the maximum, minimum, or actual size of a shaft or hole and the basic size.

Maximum metal condition (MMC). This is the maximum limit of an external feature; for example, a shaft manufactured to its high limit would contain the maximum amount of metal. It is also the minimum limit on an internal feature; for example, a component which has had a hole bored in it to its lower limit of size would have had the minimum of metal removed and remain in its maximum metal condition.

Unilateral and bilateral limits

Fig. 29.10 shows an example of unilateral limits, where the maximum and minimum limits of size are disposed on the same side of the basic size. This system is preferred since the basic size is used for the GO limit gauge; changes in the magnitude of the tolerance affect only the size of the other gauge dimension, the NOT GO gauge size.

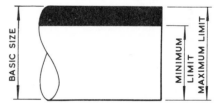

Fig. 29.10   Unilateral limits

Fig. 29.11 shows an example of bilateral limits, where the limits are disposed above and below the basic size.

Fig. 29.11   Bilateral limits

Bases of fits

The two bases of a system of limits and fits are
a) the hole basis,
b) the shaft basis.

Hole basis (fig. 29.12) In this system, the basic diameter of the hole is constant while the shaft size varies according to the type of fit. This system leads to greater economy of production, as a single drill or reamer size can be used to produce a variety of fits by merely altering the shaft limits. The shaft can be accurately produced to size by turning and grinding. Generally it is usual to recommend hole-base fits, except where temperature may have a detrimental effect on large sizes.

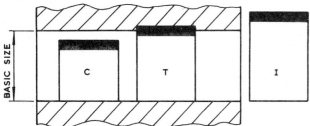

Fig. 29.12 Hole-basis fits: C - clearance T - transition
I - interference

Shaft basis (fig. 29.13) Here the hole size is varied to produce the required class of fit with a basic-size shaft. A series of drills and reamers is required for this system, therefore it tends to be costly. It may, however, be necessary to use it where different fits are required along a long shaft. The BSI data sheet 4500A gives a selection of ISO fits on the hole basis, and data sheet 4500B gives a selection of shaft-basis fits extracted from BS 4500, the current standard on limits and fits.

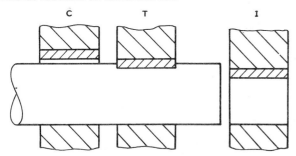

Fig. 29.13 Shaft-basis fits: C - clearance T - transition
I - interference

The ISO system contained in BS 4500 gives an extensive selection of hole and shaft tolerances to cover a wide range of applications. It has been found, however, that in the manufacture of many standard engineering components a limited selection of tolerances is adequate. These are provided on the data sheets referred to above. Obviously, by using only a selected range of fits, economic advantages are obtained from the reduced tooling and gauging facilities involved.

Selected ISO fits—hole basis (extracted from BS 4500)

The ISO system provides a great many hole and shaft tolerances so as to cater for a very wide range of conditions. However, experience shows that the majority of fit conditions required for normal engineering products can be provided by a quite limited selection of tolerances. The following selected hole and shaft tolerances have been found to be commonly applied:

  selected hole tolerances: H7 H8 H9 H11;
  selected shaft tolerances: c11 d10 e9 f7 g6 h6 k6 n6 p6 s6.

The table opposite shows a range of fits derived from these selected hole and shaft tolerances. As will be seen, it covers fits from loose clearance to heavy interference, and it may therefore be found to be suitable for most normal requirements. Many users may in fact find that their needs are met by a further selection within this selected range.

It should be noted, however, that this table is offered only as an example of how a restricted selection of fits can be made. It is clearly impossible to recommend selections of fits which are appropriate to all sections of industry, but it must be emphasised that a user who decides upon a selected range will always enjoy the economic advantages this conveys. Once he has installed the necessary tooling and gauging facilities, he can combine his selected hole and shaft tolerances in different ways without any additional investment in tools and equipment.

For example, if it is assumed that the range of fits shown in the table has been adopted but that, for a particular application the fit H8-f7 is appropriate but provides rather too much variation, the hole tolerance H7 could equally well be associated with the shaft f7 and may provide exactly what is required without necessitating any additional tooling.

For most general applications, it is usual to recommend hole-basis fits, as, except in the realm of very large sizes where the effects of temperature play a large part, it is usually considered easier to manufacture and measure the male member of a fit, and it is thus desirable to be able to allocate the larger part of the tolerance available to the hole and adjust the shaft to suit.

In some circumstances, however, it may in fact be preferable to employ a shaft basis. For example, in the case of driving shafts where a single shaft may give to accommodate a variety of accessories such as couplings, bearings, collars, etc., it is preferable to maintain a constant diameter for the permanent member, which is the shaft, and vary the bore of the accessories. For use in applications of this kind, a selection of shaft basis fits is provided in data sheet BS 4500B.

BRITISH STANDARD

SELECTED ISO FITS— HOLE BASIS

Extracted from BS 4500 : 1969

Data Sheet 4500A
Issue 1. February 1970

Diagram to scale for 25 mm. diameter

Clearance fits: H11/c11, H9/d10, H9/e9, H8/f7, H7/g6, H7/h6
Transition fits: H7/k6, H7/n6
Interference fits: H7/p6, H7/s6

All tolerance values in 0·001 mm. Each cell shows upper deviation / lower deviation.

Nominal sizes Over (mm)	To (mm)	H11	c11	H9	d10	H9	e9	H8	f7	H7	g6	H7	h6	H7	k6	H7	n6	H7	p6	H7	s6
—	3	+60 / 0	−60 / −120	+25 / 0	−20 / −60	+25 / 0	−14 / −39	+14 / 0	−6 / −16	+10 / 0	−2 / −8	+10 / 0	0 / −6	+10 / 0	+6 / 0	+10 / 0	+10 / +4	+10 / 0	+12 / +6	+10 / 0	+20 / +14
3	6	+75 / 0	−70 / −145	+30 / 0	−30 / −78	+30 / 0	−20 / −50	+18 / 0	−10 / −22	+12 / 0	−4 / −12	+12 / 0	0 / −8	+12 / 0	+9 / +1	+12 / 0	+16 / +8	+12 / 0	+20 / +12	+12 / 0	+27 / +19
6	10	+90 / 0	−80 / −170	+36 / 0	−40 / −98	+36 / 0	−25 / −61	+22 / 0	−13 / −28	+15 / 0	−5 / −14	+15 / 0	0 / −9	+15 / 0	+10 / +1	+15 / 0	+19 / +10	+15 / 0	+24 / +15	+15 / 0	+32 / +23
10	18	+110 / 0	−95 / −205	+43 / 0	−50 / −120	+43 / 0	−32 / −75	+27 / 0	−16 / −34	+18 / 0	−6 / −17	+18 / 0	0 / −11	+18 / 0	+12 / +1	+18 / 0	+23 / +12	+18 / 0	+29 / +18	+18 / 0	+39 / +28
18	30	+130 / 0	−110 / −240	+52 / 0	−65 / −149	+52 / 0	−40 / −92	+33 / 0	−20 / −41	+21 / 0	−7 / −20	+21 / 0	0 / −13	+21 / 0	+15 / +2	+21 / 0	+28 / +15	+21 / 0	+35 / +22	+21 / 0	+48 / +35
30	40	+160 / 0	−120 / −280	+62 / 0	−80 / −180	+62 / 0	−50 / −112	+39 / 0	−25 / −50	+25 / 0	−9 / −25	+25 / 0	0 / −16	+25 / 0	+18 / +2	+25 / 0	+33 / +17	+25 / 0	+42 / +26	+25 / 0	+59 / +43
40	50	+160 / 0	−130 / −290	+62 / 0	−80 / −180	+62 / 0	−50 / −112	+39 / 0	−25 / −50	+25 / 0	−9 / −25	+25 / 0	0 / −16	+25 / 0	+18 / +2	+25 / 0	+33 / +17	+25 / 0	+42 / +26	+25 / 0	+59 / +43
50	65	+190 / 0	−140 / −330	+74 / 0	−100 / −220	+74 / 0	−60 / −134	+46 / 0	−30 / −60	+30 / 0	−10 / −29	+30 / 0	0 / −19	+30 / 0	+21 / +2	+30 / 0	+39 / +20	+30 / 0	+51 / +32	+30 / 0	+72 / +53
65	80	+190 / 0	−150 / −340	+74 / 0	−100 / −220	+74 / 0	−60 / −134	+46 / 0	−30 / −60	+30 / 0	−10 / −29	+30 / 0	0 / −19	+30 / 0	+21 / +2	+30 / 0	+39 / +20	+30 / 0	+51 / +32	+30 / 0	+78 / +59
80	100	+220 / 0	−170 / −390	+87 / 0	−120 / −260	+87 / 0	−72 / −159	+54 / 0	−36 / −71	+35 / 0	−12 / −34	+35 / 0	0 / −22	+35 / 0	+25 / +3	+35 / 0	+45 / +23	+35 / 0	+59 / +37	+35 / 0	+93 / +71
100	120	+220 / 0	−180 / −400	+87 / 0	−120 / −260	+87 / 0	−72 / −159	+54 / 0	−36 / −71	+35 / 0	−12 / −34	+35 / 0	0 / −22	+35 / 0	+25 / +3	+35 / 0	+45 / +23	+35 / 0	+59 / +37	+35 / 0	+101 / +79
120	140	+250 / 0	−200 / −450	+100 / 0	−145 / −305	+100 / 0	−84 / −185	+63 / 0	−43 / −83	+40 / 0	−14 / −39	+40 / 0	0 / −25	+40 / 0	+28 / +3	+40 / 0	+52 / +27	+40 / 0	+68 / +43	+40 / 0	+117 / +92
140	160	+250 / 0	−210 / −460	+100 / 0	−145 / −305	+100 / 0	−84 / −185	+63 / 0	−43 / −83	+40 / 0	−14 / −39	+40 / 0	0 / −25	+40 / 0	+28 / +3	+40 / 0	+52 / +27	+40 / 0	+68 / +43	+40 / 0	+125 / +100
160	180	+250 / 0	−230 / −480	+100 / 0	−145 / −305	+100 / 0	−84 / −185	+63 / 0	−43 / −83	+40 / 0	−14 / −39	+40 / 0	0 / −25	+40 / 0	+28 / +3	+40 / 0	+52 / +27	+40 / 0	+68 / +43	+40 / 0	+133 / +108
180	200	+290 / 0	−240 / −530	+115 / 0	−170 / −355	+115 / 0	−100 / −215	+72 / 0	−50 / −96	+46 / 0	−15 / −44	+46 / 0	0 / −29	+46 / 0	+33 / +4	+46 / 0	+60 / +31	+46 / 0	+79 / +50	+46 / 0	+151 / +122
200	225	+290 / 0	−260 / −550	+115 / 0	−170 / −355	+115 / 0	−100 / −215	+72 / 0	−50 / −96	+46 / 0	−15 / −44	+46 / 0	0 / −29	+46 / 0	+33 / +4	+46 / 0	+60 / +31	+46 / 0	+79 / +50	+46 / 0	+159 / +130
225	250	+290 / 0	−280 / −570	+115 / 0	−170 / −355	+115 / 0	−100 / −215	+72 / 0	−50 / −96	+46 / 0	−15 / −44	+46 / 0	0 / −29	+46 / 0	+33 / +4	+46 / 0	+60 / +31	+46 / 0	+79 / +50	+46 / 0	+169 / +140
250	280	+320 / 0	−300 / −620	+130 / 0	−190 / −400	+130 / 0	−110 / −240	+81 / 0	−56 / −108	+52 / 0	−17 / −49	+52 / 0	0 / −32	+52 / 0	+36 / +4	+52 / 0	+66 / +34	+52 / 0	+88 / +56	+52 / 0	+190 / +158
280	315	+320 / 0	−330 / −650	+130 / 0	−190 / −400	+130 / 0	−110 / −240	+81 / 0	−56 / −108	+52 / 0	−17 / −49	+52 / 0	0 / −32	+52 / 0	+36 / +4	+52 / 0	+66 / +34	+52 / 0	+88 / +56	+52 / 0	+202 / +170
315	355	+360 / 0	−360 / −720	+140 / 0	−210 / −440	+140 / 0	−125 / −265	+89 / 0	−62 / −119	+57 / 0	−18 / −54	+57 / 0	0 / −36	+57 / 0	+40 / +4	+57 / 0	+73 / +37	+57 / 0	+98 / +62	+57 / 0	+226 / +190
355	400	+360 / 0	−400 / −760	+140 / 0	−210 / −440	+140 / 0	−125 / −265	+89 / 0	−62 / −119	+57 / 0	−18 / −54	+57 / 0	0 / −36	+57 / 0	+40 / +4	+57 / 0	+73 / +37	+57 / 0	+98 / +62	+57 / 0	+244 / +208
400	450	+400 / 0	−440 / −840	+155 / 0	−230 / −480	+155 / 0	−135 / −290	+97 / 0	−68 / −131	+63 / 0	−20 / −60	+63 / 0	0 / −40	+63 / 0	+45 / +5	+63 / 0	+80 / +40	+63 / 0	+108 / +68	+63 / 0	+272 / +232
450	500	+400 / 0	−480 / −880	+155 / 0	−230 / −480	+155 / 0	−135 / −290	+97 / 0	−68 / −131	+63 / 0	−20 / −60	+63 / 0	0 / −40	+63 / 0	+45 / +5	+63 / 0	+80 / +40	+63 / 0	+108 / +68	+63 / 0	+292 / +252

Holes / Shafts

Chapter 30

Geometrical tolerances

The object of this section is to illustrate and interpret in simple terms the advantages of calling for geometrical tolerances on engineering drawings, and also to show that, when correctly used, they ensure that communications between the drawing office and the workshop are complete and incapable of mis-interpretation, regardless of any language barrier.

Applications

Geometrical tolerances are applied over and above normal dimensional tolerances when it is necessary to control more precisely the form or shape of some feature of a manufactured part, because of the particular duty that the part has to perform. In the past, the desired qualities would have been obtained by adding to drawings such expressions as 'surfaces to be true with one another', 'surfaces to be square with one another', 'surfaces to be flat and parallel', etc., and leaving it to workshop tradition to provide a satisfactory interpretation of the requirements.

Advantages

Geometrical tolerances are used to convey in a brief and precise manner complete geometrical requirements on engineering drawings. They should always be considered for surfaces which come into contact with other parts, especially when close tolerances are applied to the features concerned.

No language barrier can exist, as the symbols used are in agreement with published recommendations of the International Organisation for Standardisation (ISO) and have been internationally agreed. BS 308: part 3: 1972 incorporates these symbols.

Caution It must be emphasized that geometrical tolerances should be applied only when real advantages result, when normal methods of dimensioning are considered inadequate to ensure that the design function is kept, especially where repeatability must be guaranteed. Indiscriminate use of geometrical tolerances could increase costs in manufacture and inspection. Tolerances should be as wide as possible, as the satisfactory design function permits.

General rules

The symbols relating to geometrical tolerances are shown in Table 30.1. Examination of the various terms—flatness, straightness, concentricity, etc.—used to describe the geometrical characteristics shows that one type of geometrical tolerance can control another form of geometrical error. For example, a positional tolerance can control squareness and straightness; parallelism, squareness, and angularity tolerances can control flatness.

The use of geometrical tolerances does not involve or imply any particular method of manufacture or inspection.

Geometrical tolerances shown in this book, and in keeping with European conventions, must be met regardless of feature size, unless modified by Ⓜ–maximum material condition.

Maximum material condition, denoted by the symbol Ⓜ describes a part which contains the maximum amount of material, i.e. the minimum size hole or the maximum size shaft.

Boxed dimensions indicate true position, which is the theoretical exact location of features such as holes, slots, bosses, etc. Boxed dimensions are never individually toleranced but must always be accompanied by a positional or zone tolerance for the feature to which they refer.

Definitions

Limits. The maximum and minimum dimensions for a given feature are known as the 'limits'.

For example, 20 ± 0.1 or $\begin{matrix} 20.1 \\ 19.9 \end{matrix}$

The upper and lower limits of size are 20.1mm and 19.9mm respectively.

Tolerance. The algebraic difference between the upper and lower limit of size is known as the 'tolerance'. In the example above, the tolerance is 0.2mm. The tolerance is the amount of variation permitted.

Nominal dimension. Limits and tolerances are based on 'nominal dimensions' which are target dimensions. In practice there is no such thing as a nominal dimension, since no part can be machined to an exact size, but the use of the term is an accepted convention.

The limits referred to above can be set in two ways:
a) *unilateral limits*–limits set wholly above or below the nominal size;
b) *bilateral limits*–limits set partly above and partly below the nominal size.

Geometrical tolerance. These tolerances specify the maximum error of a component's geometrical shape, position, or size over its whole length or surface. This is done by defining a zone in which the feature may lie, and this may be modified only by a more restrictive explanatory note.

Tolerance zone. A tolerance zone is an area or volume, enclosed by imaginary lines or surfaces, in which any deviation of the feature must be contained.

Method of indicating geometrical tolerances on drawings

Geometrical tolerances are indicated by stating the follow-

Tolerance characteristics			*Symbol*	*Applications*
Single features	Form tolerances:	Straightness	—	A straight line. The edge or axis of a feature.
		Flatness	▱	A plane surface.
		Roundness	◯	The periphery of a circle. Cross-section of a bore, cylinder, cone, or sphere.
		Cylindricity	⌭	The combination of roundness, straightness, and parallelism of cylindrical surfaces. Mating bores and plungers.
		Profile of a line	⌒	The theoretical or perfect form of profiles defined by true boxed dimensions.
		Profile of a surface	⌓	The theoretical or perfect form of surfaces defined by true boxed dimensions.
Related features	Attitude tolerances:	Parallelism	//	Parallelism of a feature related to a datum. Can control flatness when related to a datum.
		Squareness	⊥	Surfaces, axes, or lines positioned at right angles to each other.
		Angularity	∠	The angular displacement of surfaces, axes, or lines from a datum.
	Composite tolerances:	Runout	↗	The position of a point fixed on a surface of a part which is rotated 360° about its axis.
	Location tolerances:	Position	⊕	The deviation of a feature from a true position.
		Concentricity	◎	The relationship between two cylinders or circles having a common axis.
		Symmetry	⩵	The symmetrical position of a feature related to a datum.

Table 30.1 Geometrical-tolerance characteristics with their identification symbols and applications

ing details in compartments in a rectangular frame:
a) the characteristic symbol, for single or related features;
b) the tolerance value,
   i) preceded by $\emptyset$ if the zone is circular or cylindrical.
   ii) preceded by SPHERE $\emptyset$ if the zone is spherical;
c) letters identifying datum features when specified.
Fig. 30.1 shows examples.

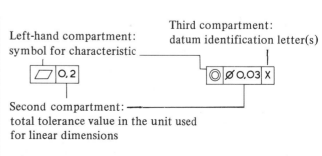

Left-hand compartment: symbol for characteristic

Third compartment: datum identification letter(s)

Second compartment: total tolerance value in the unit used for linear dimensions

Fig. 30.1

Methods of applying the tolerance frame to the toleranced feature

Figs 30.2 and 30.3 illustrate alternative methods of referring the tolerance to the surface or the plane itself. Note that in fig. 30.3 the dimension line and frame leader line are offset.

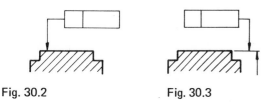

Fig. 30.2   Fig. 30.3

Figs 30.4 and 30.5 illustrate alternative methods of referring the tolerance to the axis or median plane. Note that in fig. 30.5 the dimension line and frame leader line are drawn in line.

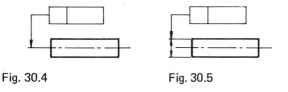

Fig. 30.4   Fig. 30.5

The tolerance frame as shown in fig. 30.6 refers to the axis or median plane of the 'whole' part.

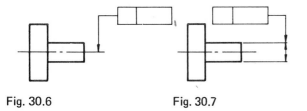

Fig. 30.6   Fig. 30.7

The tolerance frame as shown in fig. 30.7 refers to the axis or median plane only of the dimensioned feature.

The application of tolerances to a restricted length of a feature

Fig. 30.8 shows the method of applying a tolerance to only a particular part of a feature.

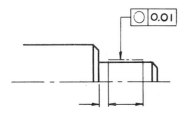

Fig. 30.8

The tolerance frame in fig. 30.9 shows the method of applying another tolerance, similar in type but smaller in magnitude, on a shorter length. In this case, the whole flat surface must lie between parallel planes 0.2 apart, but over any length of 180 mm, in any direction, the surface must be within 0.05.

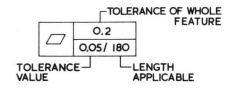

Fig. 30.9

Fig. 30.10 shows the method used to apply a tolerance over a given length; it allows the tolerance to accumulate over a longer length. In this case, a parallel tolerance of 0.02 is applicable over a length of 100mm. If the total length of the feature were 800mm, then the total permitted tolerance would accumulate to 0.16.

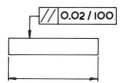

Fig. 30.10

Maximum material condition

The symbol for maximum material condition is the letter M enclosed by a circle, Ⓜ. The symbol is positioned in the tolerance frame as follows:

a) $\boxed{\oplus \;|\; \varnothing\; 0.05\text{Ⓜ} \;|\; X}$ Refers to the tolerance only.

b) $\boxed{\oplus \;|\; \varnothing 0.05 \;|\; X\text{Ⓜ}}$ Refers to the datum only.

c) $\boxed{\oplus \;|\; \varnothing 0.05\text{Ⓜ} \;|\; X\text{Ⓜ}}$ Refers to both tolerance and datum.

Chapter 31

Datums

Datum

A datum is in practice the functional plane or axis used for manufacture or inspection purposes, and should always be clearly identified on the drawing. A datum surface on a component should be accurately finished, since other locations or surfaces are established by measuring from the datum. Fig. 31.1 shows a datum surface indicated by the letter A.

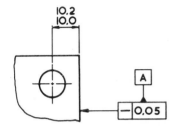

Fig. 31.1

In the above example, the datum edge is subject to a straightness tolerance of 0.05, shown in the tolerance frame. This datum is given a letter, to distinguish it from other datums, and the letter is boxed and connected to the tolerance frame by a leader line terminating in a solid equilateral triangle.

Methods of specifying datum features

Single datums

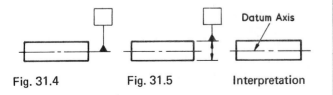

Fig. 31.2   Fig. 31.3   Interpretation

Figs, 31.2 and 31.3 illustrate alternative methods of positioning the datum box when the surface itself is the datum. Note that in fig. 31.3 the dimension line and datum box are offset.

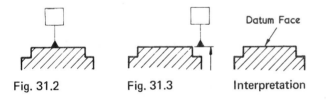

Fig. 31.4   Fig. 31.5   Interpretation

Figs, 31.4 and 31.5 illustrate alternative methods of positioning the datum box when the axis or median plane is the datum. Note that in fig. 31.5 the dimension line and the datum box are in line.

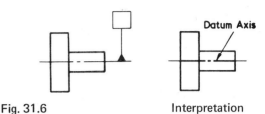

Fig. 31.6   Interpretation

The datum box shown in Fig. 31.6 indicates that the whole length of the common axis or median plane of two or more features is the datum.

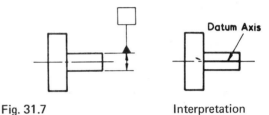

Fig. 31.7   Interpretation

The datum box in fig. 31.7 indicates that the datum axis or median plane refers only to the feature dimensioned.

Multiple datums

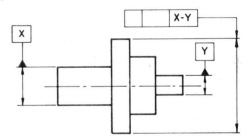

Fig. 31.8

Fig. 31.8 illustrates two datum features of equal status used to establish a single datum plane. The reference letters are placed in the third compartment of the tolerance frame, and have a hyphen separating them.

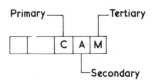

Fig. 31.9

Fig. 31.9 shows the drawing instruction necessary when the application of the datum features is required in a particular order of priority.

111

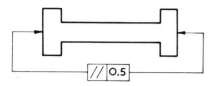

Fig. 31.10

Fig. 31.10 illustrates the case where there is no functional reason to choose one or other of two associated features as a datum. The tolerance frame is applied as shown.

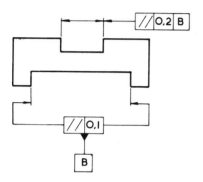

Fig. 31.11

Fig. 31.11 illustrates the case where two associated features are chosen as a datum, from which another feature is toleranced.

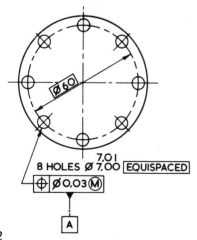

Fig. 31.12

Fig. 31.12 shows a datum for a positional tolerance which is derived from the true position of a group of holes.

Fig. 31.13 shows an application where a geometrical tolerance is related to two separate datum surfaces.

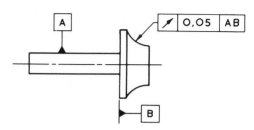

Fig. 31.13

Functionally, both datums are of equal importance, and no priority is implied. Additional letters can be added in the third compartment if further datum references are required. No significance is attached to the letters used.

If a positional tolerance is required as shown in fig. 31.14, and no particular datum is specified, then the individual feature to which the geometrical-tolerance frame is connected is chosen as the datum.

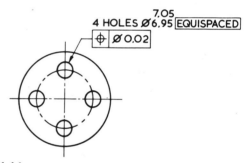

Fig. 31.14

Fig. 31.15 illustrates a further positional-tolerance application where the geometrical reference frame is to be located in relation to another feature which may be indicated as the datum. In the given example, this implies that the pitch circle and the datum circle must be concentric, i.e. they have common axes.

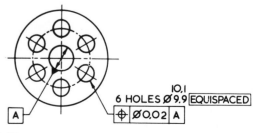

Fig. 31.15

Chapter 32

Application of geometrical tolerances

In this chapter, examples are given of the application of tolerances to each of the characteristics on engineering drawings by providing a typical specification for the product and the appropriate note which must be added to the drawing. In every example, the tolerance values given are only typical figures; the product designer would normally be responsible for the selection of tolerance values for individual cases.

Straightness
A straight line is the shortest distance between two points. A straightness tolerance controls:
1 the straightness of a line on a surface,
2 the straightness of a line in a single plane,
3 the straightness of an axis.

Case 1
Product requirement
The specified line shown on the surface must lie between two parallel straight lines 0.03 apart.

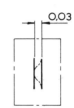

Drawing instruction
A typical application for this type of tolerance could be a graduation line on an engraved scale.

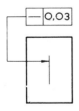

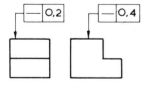

In the application shown above, tolerances are given controlling the straightness of two lines at right angles to one another. In the left-hand view the straightness control is 0.2, and in the right-hand view 0.4. As in the previous example, the position of the graduation marks would be required to be detailed on a plan view.

Case 2
Product requirement
The axis of the whole part must lie in a boxed zone of 0.3 x 0.2 over its length.

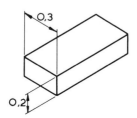

Drawing instruction
As indicated, the straightness of the axis is controlled by the dimensions of the box, and could be applied to a long rectangular key.

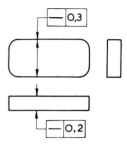

Case 3
Product requirement
The axis of the whole feature must lie within the cylindrical tolerance zone of 0.05.

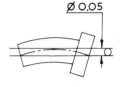

Drawing instruction

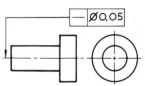

Case 4
Product requirement
The geometrical tolerance may be required to control only part of the component. In this example the axis of the dimensioned portion of the feature must lie within the cylindrical tolerance zone of 0.1 diameter.

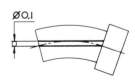

Drawing instruction

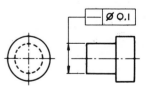

Flatness

Flatness tolerances control the divergence or departure of a surface from a true plane.

The tolerance of flatness is the specified zone between two parallel planes. It does not control the squareness or parallelism of the surface in relation to other features, and it can be called for independently of any size tolerance.

Case 1

Product requirement
The surface must be contained between two parallel planes 0.07 apart.

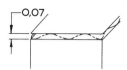

Drawing instruction

Case 2

Product requirement
The surface must be contained between two parallel planes 0.03 apart, but must not be convex.

Drawing instruction
Note that these instructions could be arranged to avoid a concave condition.

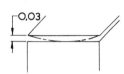

Case 3

Product requirement
The surface must be contained between two parallel planes 0.05 apart.

Drawing instruction
The example shows a control for the flatness of the seating on a special-purpose bolt.

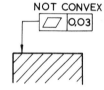

Roundness

Roundness is a condition where any point of a feature's continuous curved surface is equidistant from its centre, which lies in the same plane.

The tolerance of roundness controls the divergence of the feature, and the annular space between the two coplanar concentric circles defines the tolerance zone, the magnitude being the algebraic difference of the radii of the circles.

For a cone or cylinder, the errors of form are contained within two concentric circles which are in a plane perpendicular to the axis. For a sphere, the maximum diameter is the position of the intersection plane, unless otherwise stated. When controlling the roundness of a sphere, the word 'SPHERE' should appear above the tolerance frame containing the roundness symbol and tolerance value.

Case 1

Product requirement
The circumference of the bar must lie between two co-planer concentric circles 0.5 apart.

Drawing instruction

Note that, at any particular section, a circle may not be concentric with its axis but may still satisfy a roundness tolerance. The following diagram shows a typical condition.

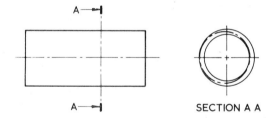

SECTION A A

Case 2

Product requirement
The circumference at any cross section must lie between two coplanar concentric circles 0.02 apart.

Drawing instruction

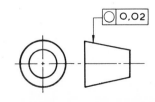

Case 3

Product requirement
The periphery of any section of maximum diameter of the sphere must lie between concentric circles a radial distance 0.001 apart in the plane of the section.

Drawing instruction

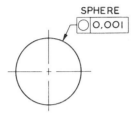

Cylindricity

The combination of parallelism, roundness, and straightness defines cylindricity when applied to the surface of a cylinder, and is controlled by a tolerance of cylindricity. The tolerance zone is the annular space between two coaxial cylinders, the radial difference being the tolerance value to be specified.

It should be mentioned that, due to difficulties in checking the combined effects of parallelism, roundness, and straightness, it is recommended that each of these characteristics is toleranced and inspected separately. The ends of the cylinder are also not covered by this standard, and, if these are to be controlled, then a limit of length is required, and squareness tolerances must be applied.

Product requirement
The whole curved surface of the feature must lie between an annular tolerance zone 0.04 wide formed by two cylindrical surfaces coaxial with each other.

Drawing instruction

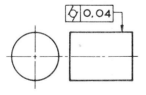

Profile tolerance of a line

Profile tolerance of a line is used to control the ideal contour of a feature. The contour is defined by boxed true-position dimensions and must be accompanied by a

relative tolerance zone. This tolerance zone, unless otherwise stated, is taken to be equally disposed about the true form, and the tolerance value is equal to the diameter of circles whose centres lie on the true form. If it is required to call for the tolerance zone to be positioned on one side of the true form (i.e. unilaterally), the circumferences of the tolerance-zone circles must touch the theoretical contour of the profile.

Case 1 *Product requirement*
The profile is required to be contained within the bilateral tolerance zone.

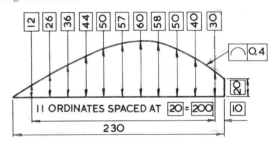

Drawing instruction

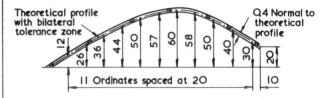

Case 2
Product requirement
The profile is required to be contained within the unilateral tolerance zone.

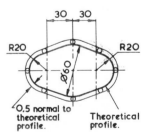

Drawing instruction

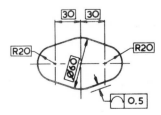

The figure below shows an example where the toleranced profile of a feature has a sharp corner. The inner tolerance zone is considered to end at the intersection of the inner boundary lines, and the outer tolerance zone is considered to extend to the outer boundary-line intersections. Sharp corners such as these could allow considerable rounding; if this is desirable, then the design requirement must be clearly defined on the drawing by specifying a radius or adding a general note such as 'ALL CORNERS 0.5 MAX'. It should be noted that such radii apply regardless of the profile tolerance, and they are called for as true-position boxed dimensions.

In the example given, the product is required to have a sharp corner.

Product requirement

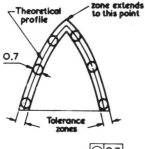

Drawing instruction

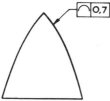

Profile tolerance of a surface

Profile tolerance of a surface is used to control the ideal form of a surface, which is defined by true-position boxed dimensions and must be accompanied by a relative tolerance zone. The profile-tolerance zone, unless otherwise stated, is taken to be bilateral and equally disposed about its true-form surface. The tolerance value is equal to the diameter of spheres whose centre lines lie on the true form of the surface. The zone is formed by surfaces which touch the circumferences of the spheres on either side of the ideal form.

If it is required to call for a unilateral tolerance zone, then the circumferences of the tolerance-zone spheres must touch the theoretical contour of the surface.

Product requirement
The tolerance zone is to be contained by upper and lower surfaces which touch the circumference of spheres 0.3 diameter whose centres lie on the theoretical form of the surface.

Drawing instruction

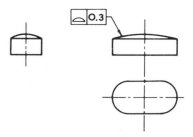

Parallelism

Two parallel lines or surfaces are always separated by a uniform distance. Lines or surfaces may be required to be parallel with datum planes or axes. Tolerance zones may be the space between two parallel lines or surfaces, or the space contained within a cylinder positioned parallel to its datum. The magnitude of the tolerance value is the distance between the parallel lines or surfaces, or the cylinder diameter.

Case 1
Product requirement
The axis of the hole on the left-hand side must be contained between two straight lines 0.2 apart, parallel to the datum axis X and lying in the same vertical plane.

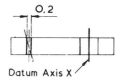

Drawing instruction

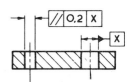

Case 2
Product requirement
The axis of the upper hole must be contained between two straight lines 0.3 apart which are parallel to the datum axis X and lie in the same horizontal plane.

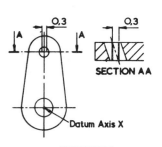

Drawing instruction

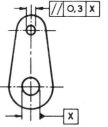

Case 3

Product requirement

The upper hole axis must be contained in a cylindrical zone 0.4 diameter, with its axis parallel to the datum X.

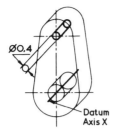

Drawing instruction

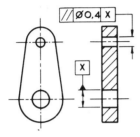

Case 4

Product requirement

The axis of the hole on the left-hand side must be contained in a tolerance box 0.5 x 0.2 x width, as shown, with its sides parallel to the datum axis X and in the same horizontal plane.

Drawing instruction

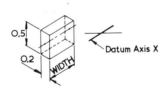

Case 5

Product requirement

The axis of the hole must be contained between two planes 0.06 apart parallel to the datum surface X.

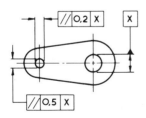

Drawing instruction

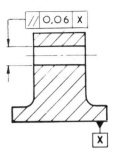

Case 6

Product requirement

The top surface of the component must be contained between two planes 0.7 apart and parallel to the datum X.

Drawing instruction

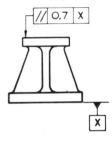

Squareness (perpendicularity)

Squareness is the condition when a line, plane, or surface is at right angles to a datum feature.

The tolerance zone is the space between two parallel lines or surfaces; it can also be the space contained within a cylinder. All tolerance zones are perpendicular to the datum feature.

The magnitude of the tolerance value is the specified distance between these parallel lines or surfaces, or the diameter of the cylinder.

Case 1

Product requirement

The axis of the vertical hole must be contained between two planes 0.1 apart which are perpendicular to the datum axis.

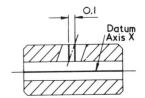

Drawing instruction

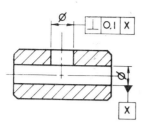

Case 2

Product requirement

The axis of the upright must be contained between two straight lines 0.2 apart which are perpendicular to the datum. Squareness is controlled here in one plane only.

Drawing instruction

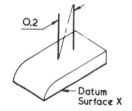

Case 3

Product requirement

The axis of the column must be contained in a cylindrical tolerance zone 0.3 diameter, the axis of which is perpendicular to the datum surface X. Squareness is controlled in more than one plane by this method.

Drawing instruction

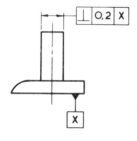

Case 4

Product requirement

The axis of the column must be contained in a tolerance-zone box 0.2 x 0.4 which is perpendicular to the datum surface X.

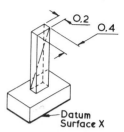

Drawing instruction

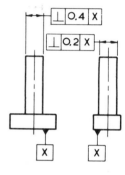

Case 5

Product requirement

The left-hand end face of the part must be contained between two parallel planes 0.8 apart and perpendicular to the datum axis X.

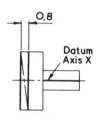

Drawing instruction

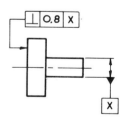

Case 6

Product requirement

The left-hand surface must be contained between two parallel planes 0.7 apart and perpendicular to the datum surface X.

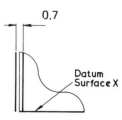

Drawing instruction

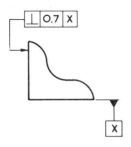

Angularity

Angularity defines a condition between two related planes, surfaces, or lines which are not perpendicular or parallel to one another. Angularity tolerances control this relationship.

The specified angle is a basic dimension, and is defined by a true-position boxed dimension and must be accompanied by a tolerance zone. This zone is the area between two parallel lines inclined at the specified angle to the datum line, plane, or axis. The tolerance zone may also be the space within a cylinder, the tolerance value being equal to the cylinder diameter; In this case, symbol ⌀ precedes tolerance value in the tolerance frame.

Case 1

Product requirement
The inclined surface must be contained within two parallel planes 0.2 apart which are at an angle of 42° to the datum surface.

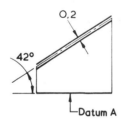

Drawing instruction

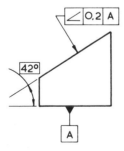

Case 2

Product requirement
The axis of the hole must be contained within two parallel straight lines 0.1 apart inclined at 28° to the datum axis.

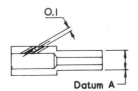

Drawing instruction

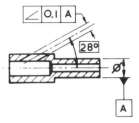

Case 3

Product requirement
The axis of the inclined hole must be within a cylindrical tolerance zone of 0.3 diameter inclined at 45° to the datum plane.

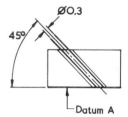

Drawing instruction

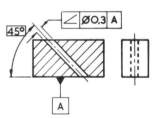

Case 4

Product requirement
The inclined surface must be contained within two parallel planes 0.5 apart which are inclined at 100° to the datum axis.

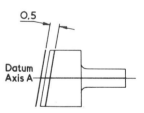

Drawing instruction

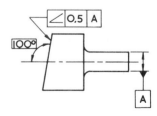

Run-out

Run-out is a unique geometrical tolerance. It can be a composite form control relating two or more characteristics, and it requires a practical test where the part is rotated through 360° about its own axis.

The results of this test may include errors of other characteristics such as roundness, concentricity, perpendicularity, or flatness, but the test cannot discriminate between them. It should therefore not be called for where the design function of the part necessitates that the other characteristics are to be controlled separately. The total of any of these errors will be contained within the specified run-out tolerance value. The tolerance value is equal to the full indicator movement of a fixed point measured during one revolution of the component about its axis, without axial movement. Run-out is measured in the direction specified by the arrow at the end of the leader line which points to the toleranced feature. It must always be measured regardless of feature size, and it is convenient for practical purposes to establish the datum as the diameter(s) from which measurement is to be made, if true centres have not been utilised in manufacturing the part.

Case 1
Product requirement
Run-out must not exceed 0.4 at any point along the cylinder, measured perpendicular to the datum axis without axial movement.

Drawing instruction

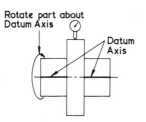

Case 2
Product requirement
Run-out must not exceed 0.4 at any point along the cylinder, measured perpendicular to the datum diameters without axial movement.

Drawing instruction

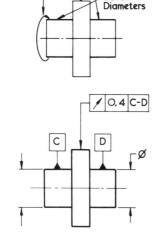

Case 3
Product requirement
Run-out must not exceed 0.2 measured at any point normal to the surface, without axial movement.

Drawing instruction

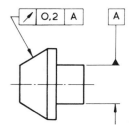

Case 4
Product requirement
At any radius, the run-out must not exceed 0.06 measured parallel to the datum axis.

Drawing instruction

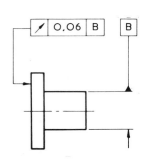

Case 5 *Product requirement*
The component is required to be rotated about datum axis C, with datum face B set to ensure no axial movement.

Run-out on the cylindrical portion must not exceed 0.05 at any point measured perpendicular to the datum axis.

Run-out on the tapered portion must not exceed 0.07 at any point measured normal to its surface.

Run-out on the curved portion must not exceed 0.04 at any point measured normal to its surface.

Run-out of the end face must not exceed 0.1 at any point measured parallel to the datum axis of rotation.

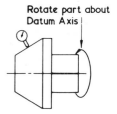

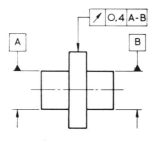

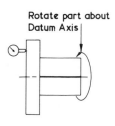

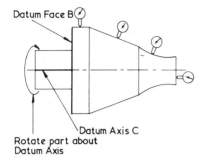

Datum Face B

Datum Axis C

Rotate part about
Datum Axis

Drawing Instruction

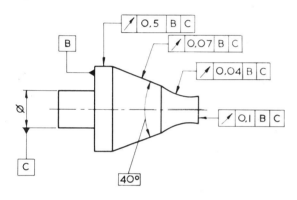

/0,5	B	C
/0,07	B	C
/0,04	B	C
/0,1	B	C

B

C

Ø

40°

Position

A positional tolerance controls the location of one feature from another feature or datum.

The tolerance zone can be the space between two parallel lines or planes, a circle, or a cylinder. The zone defines the permissible deviation of a specified feature from a true position.

The tolerance value is the distance between the parallel lines or planes, or the diameters of the circle or cylinder.

True position also incorporates squareness and parallelism of the tolerance zones with the plane of the drawing.

Case 1
Product requirement

The point must be contained within a circle of 0.1 diameter in the plane of the surface. The circle has its centre at the intersection of the two true-position dimensions. If the point were to be located by three dimensions, the tolerance zone would be a sphere.

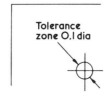

Tolerance
zone 0.1 dia

Drawing instruction

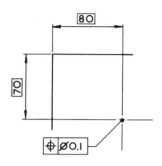

80

70

⊕ | Ø0.1

Case 2
Product requirement

The axis of the hole must be contained in a cylindrical tolerance zone of 0.3 diameter with its axis coincident with the true position of the hole axis.

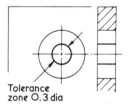

Tolerance
zone 0.3 dia

Drawing instruction

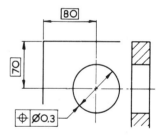

80

70

⊕ | Ø0.3

Case 3
Product requirement

Each line must be contained between two parallel straight lines on the surface, 0.2 apart, which are symmetrical with the true positions of the required lines.

Drawing instruction

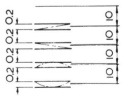

⊕ | 0.2 | 4 LINES

Case 4
Product requirement

The axes of each of the four holes must be contained in a cylindrical tolerance zone of 0.5 diameter, with its own axis coincident with the true position of each hole.

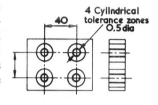

Drawing instruction

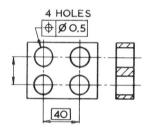

Case 5
Product requirement

The axes of each of the four holes must be contained in a boxed zone of 0.04 x 0.03 x 10, symmetrically disposed about the true position of each hole axis.

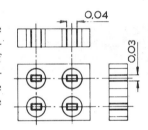

Drawing instruction

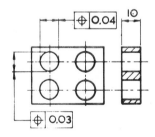

Case 6
Product requirement

The angled surface must be contained between two parallel planes 0.7 apart, which are symmetrically disposed about the true position of the surface relative to datum axis X and datum face Y.

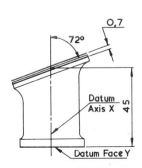

Drawing instruction

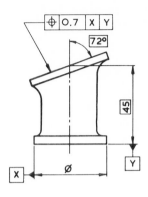

Case 7
Product requirement

The axes of the two holes must be contained in cylindrical tolerance zones of 0.01 diameter, with their own axes coincident with the true hole positions related to datum face X and the datum centre-line axis Y.

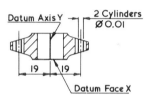

Drawing instruction

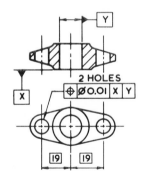

Concentricity

Two circles or cylinders are said to be concentric when their centres are coincident or they have the same axis. The deviation from the true centre or datum axis is controlled by the magnitude of the tolerance zone.

Case 1
Product requirement

To contain the centre of the large circle within a circular tolerance zone of 0.001 diameter which has its centre coincident with the datum-circle centre.

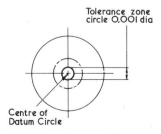

Drawing instructions

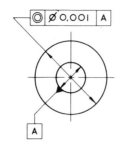

Drawing instruction

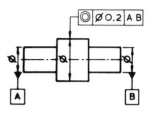

Case 2
Product requirement

To contain the axis of the right-hand cylinder within a cylindrical tolerance zone which is concentric with the axis of the datum cylinder.

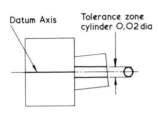

Drawing instruction

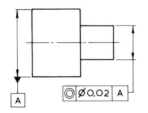

Symmetry

Symmetry involves the division of spacing of features so that they are positioned equally in relation to a datum line or plane. The tolerance zone is the space between two parallel lines or planes, parallel to, and positioned symmetrically with, the datum. The tolerance magnitude is the distance between these two parallel lines or planes.

Case 1
Product requirement

The specified line XX must lie in a tolerance zone formed by two parallel straight lines 0.01 apart and disposed symmetrically between datums A and B.

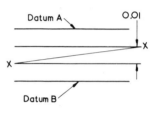

Case 3
Product requirement

To contain the axes of both the left- and right-hand cylinders within a cylindrical tolerance zone which is concentric with the axis of the datum cylinder in the centre.

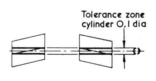

Drawing instruction

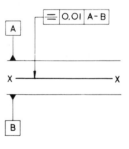

Drawing instruction

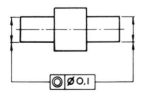

Case 4
Product requirement

To contain the axis of the central cylinder within a cylindrical tolerance zone which is coaxial with the mean axes of the left- and right-hand cylinders.

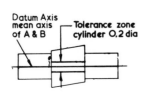

Case 2
Product requirement

The axis of the hole must lie in a tolerance zone formed by two straight parallel lines 0.02 apart, disposed symmetrically about the common median plane formed by the datum tongues X and Y.

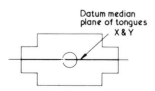

Drawing instruction

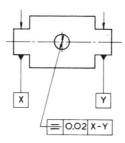

Case 3

Product requirement

The axis of a hole in a plate must lie in a rectangular-box tolerance zone 0.03 x 0.06 x depth of the plate, parallel with and symmetrically disposed about the common median planes formed by slots AC and BD.

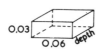

Drawing instruction

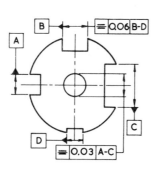

Case 4

Product requirement

The median plane of the part dimensioned X must be contained between two parallel planes 0.2 apart, symmetrically disposed about the median plane of the part dimensioned Y.

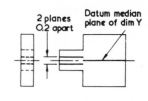

Drawing instruction

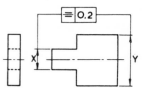

Case 5

Product requirement

The median plane of the slot and tongue must lie between two parallel planes 0.4 apart.

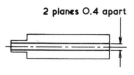

Drawing instruction

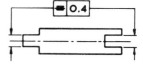

Chapter 33

Maximum material condition

Maximum material condition (denoted by the symbol ⓜ) is the state where a feature contains the maximum amount of material, i.e. the low limit of size of an internal feature and the upper limit of size of an external feature.

Any errors of form or position between two mating parts have the effect of virtually altering their respective sizes. The tightest condition of assembly between two mating parts occurs when each feature is at the MMC, plus the maximum errors permitted by any required geometrical tolerance.

Assembly clearance is increased if the actual sizes of the mating features are finished away from their MMC, and if any errors of form or position are less than that called for by any geometrical control. Also, if either part is finished away from its MMC, the clearance gained could allow for an increased error of form or position to be accepted. Any increase of tolerance gained this way, provided it is functionally acceptable for the design, is advantageous during the process of manufacture.

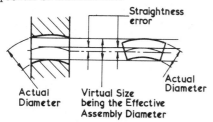

Fig. 33.1

Maximum material condition related to geometrical form

The limit of size, together with geometrical form or position of a feature, are factors of the maximum material principle, and its application is restricted to those features whose size is specified by toleranced dimensions incorporating an axis or median plane. It can never be applied to a plane, surface, or line on a surface.

The characteristics to which it can be applied are as follows:

> straightness,
> parallelism,
> squareness,
> angularity,
> position,
> concentricity,
> symmetry.

The characteristics to which the maximum material condition concept cannot be applied are as follows:

> flatness,
> roundness,
> cylindricity.
> profile of a line,
> profile of a surface,
> run-out.

Maximum material condition applied to straightness

Fig. 33.2 shows a typical drawing instruction where limits of size are applied to a pin, and in addition a straightness tolerance of 0.2 is applicable at the maximum material condition.

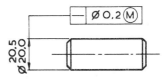

Fig. 33.2

Fig. 33.3 shows the condition where the pin is finished at the maximum material condition with the maximum straightness error. The effective assembly diameter will be equal to the sum of the upper limit of size and the straightness tolerance.

The straightness error is contained within a cylindrical tolerance zone of ∅0.2.

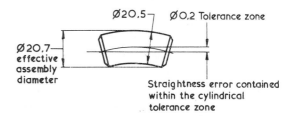

Fig. 33.3

To provide the same assembly diameter of 20.7 as shown in fig. 33.4 when the pin is finished at its low limit of size of 20.0, it follows that a straightness error of 0.7 could be acceptable. This increase may in some cases have no serious effect on the function of the component, and can be permitted.

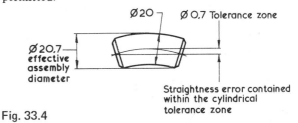

Fig. 33.4

Maximum material condition applied to squareness

Fig. 33.5 shows a typical drawing instruction where limits of size are applied to a pin, and in addition a squareness tolerance of 0.3 is applicable at the maximum material condition.

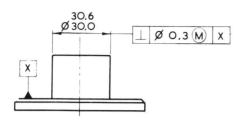

Fig. 33.5

Fig. 33.6 shows the condition where the pin is finished at the maximum material condition with the maximum squareness error of 0.3. The effective assembly diameter will be the sum of the upper limit of size and the squareness error. The squareness error will be contained within a cylindrical tolerance zone of Ø0.3.

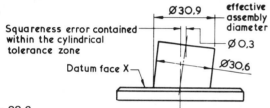

Fig. 33.6

To provide the same assembly diameter of 30.9, as shown in fig. 33.7, when the pin is finished at its low limit of size of 30.0, it follows that the squareness error could increase from 0.3 to 0.9. This permitted increase should be checked for acceptability.

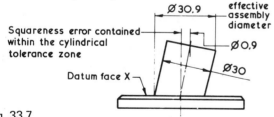

Fig. 33.7

Maximum material condition applied to position

A typical drawing instruction is given in fig. 33.8, and the following illustrations show the various extreme dimensions which can possibly arise.

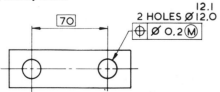

Fig. 33.8

Condition A (fig. 33.9)
Minimum distance between hole centres and the maximum material condition of holes

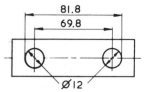

Fig. 33.9

Condition B (fig. 33.10)
Maximum distance between hole centres and maximum material condition of holes

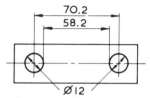

Fig. 33.10

Condition C (fig. 33.11)
To give the same assembly condition as in A, the minimum distance between hole centres is reduced when the holes are finished away from the maximum material condition.

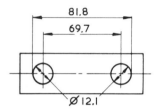

Fig. 33.11

Condition D (fig. 33.12)
To give the same assembly condition as in B, the maximum distance between hole centres is increased when the holes are finished away from the maximum material condition.

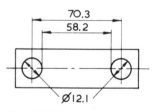

Fig. 33.12

Note that the total tolerance zone is 0.2 + 0.1 = 0.3, and therefore the positional tolerance can be increased where the two holes have a finished size away from the maximum material condition.

Maximum material condition applied to concentricity

In the previous examples, the geometrical tolerance has been related to a feature at its maximum material condition, and, provided the design function permits, the tolerance has increased when the feature has been finished away from the maximum material condition. Now the geometrical tolerance can also be specified in relation to a datum feature, and fig. 33.13 shows a typical application and drawing instruction of a shoulder on a shaft. The shoulder is required to be concentric with the shaft, which

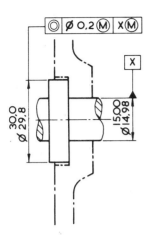

Fig. 33.13

acts as the datum. Again, provided the design function permits, further relaxation of the quoted geometrical control can be achieved by applying the maximum material condition to the datum itself.

Various extreme combinations of size for the shoulder and shaft can arise, and these are given in the drawings below. Note that the increase in concentricity error which could be permitted in these circumstances is equal to the total amount that the part is finished away from its maximum material condition, i.e. the shoulder tolerance plus the shaft tolerance.

Condition A (fig. 33.14). Shoulder and shaft at maximum material condition; shoulder at maximum permissible eccentricity to the shaft datum axis X.

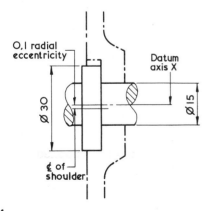

Fig. 33.14

Condition B (fig. 33.15). Shoulder at minimum material condition and shaft at maximum material condition. Total concentricity tolerance = specified concentricity tolerance + limit of size tolerance of shoulder = 0.2 + 0.2 = 0.4 diameter. This gives a maximum eccentricity of 0.2.

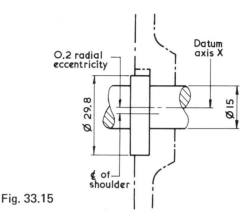

Fig. 33.15

Condition C. Fig. 33.16 shows the situation where the smallest size shoulder is associated with the datum shaft at its low limit of size. Here the total concentricity tolerance which may be permitted is the sum of the specified concentricity tolerance + limit of size tolerance for the shoulder + tolerance on the shaft = 0.2 + 0.2 + 9.02 = 0.42 diameter.

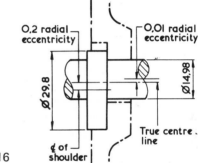

Fig. 33.16

Maximum material condition and perfect form

When any errors of geometrical form are required to be contained within the maximum material limits of size, it is assumed that the part will be perfect in form at the upper limit of size.

In applying this principle, two conditions are possible.

Case 1. The value of the geometrical tolerance can progressively increase provided that the assembly diameter does not increase above the maximum material limit of size. Fig. 33.17 shows a shaft and the boxed dimension, and indicates that at maximum material limit of size the shaft is required to be perfectly straight.

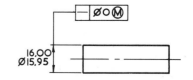

Fig. 33.17

Fig. 33.18 shows the shaft manufactured to its lower limit of size, where the permitted error in straightness can be 0.05, since the assembly diameter will be maintained at 16.00. Similarly, a shaft manufactured to, say, 15.97 can have a permitted straightness error of 0.03.

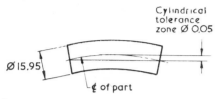

Fig. 33.18

Case 2. The geometrical tolerance can also be limited to a certain amount where it would be undesirable for the part to be used in service too much out of line.

Fig. 33.19 shows a shaft, and the boxed dimension indicates that at the maximum material limit of size the shaft is required to be perfectly straight. Also, the upper part of the box indicates that a maximum geometrical tolerance error of 0.02 can exist, provided that for assembly purposes the assembly diameter does not exceed 14.00.

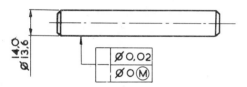

Fig. 33.19

Fig. 33.20 shows the largest diameter shaft acceptable, assuming that it has the full geometrical error of 0.02. Note that a shaft finished at 13.99 would be permitted a maximum straightness error of only 0.01 to conform with the drawing specification.

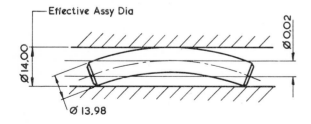

Fig. 33.20

Fig. 33.21 shows the smallest diameter shaft acceptable, and the effect of the full geometrical error of straightness.

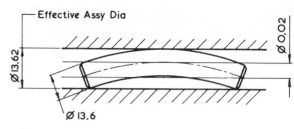

Fig. 33.21

The application of maximum material condition and its relationship with perfect form and squareness

A typical drawing instruction is shown in fig. 33.22.

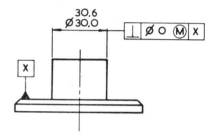

Fig. 33.22

Condition A (fig. 33.23). Maximum size of feature; zero geometrical tolerance.

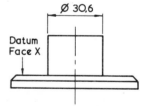

Fig. 33.23

Condition B (fig. 33.24). Minimum size of feature; Permitted geometrical error = 0.6.

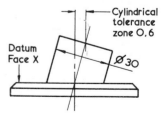

Fig. 33.24

Note that between these extremes the geometrical tolerance will progressively increase; i.e. when the shaft diameter is 30.3, then the cylindrical tolerance error permitted will be 0.3.

The application of maximum material condition and its relationship with perfect form and concentricity

A typical drawing instruction is shown in fig. 33.25.

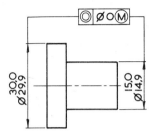

Fig. 33.25

Condition A (fig. 33.26). Head and shank at maximum material condition. No geometrical error is permitted, and the two parts of the component are concentric.

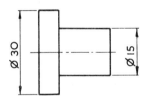

Fig. 33.26

Condition B (fig. 33.27). Head at maximum material condition; shank at minimum material condition. The permitted geometrical error is equal to the tolerance on the shank size. This gives a tolerance zone of 0.1 diameter.

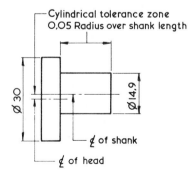

Fig. 33.27

Condition C (fig. 33.28). Shank at maximum material condition; head at minimum material condition. The permitted geometrical error is equal to the tolerance on the head size. This gives a tolerance zone of 0.1 diameter.

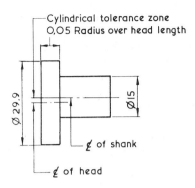

Fig. 33.28

Condition D (fig. 33.29). Both shank and shaft are finished at their low limits of size; hence the permitted geometrical error will be the sum of the two manufacturing tolerances, namely 0.2 diameter.

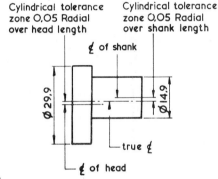

Fig. 33.29

The application of maximum material condition to two mating components

Fig. 33.30 shows a male and female component dimensioned with a linear tolerance between centres, and which will assemble together under the most adverse conditions

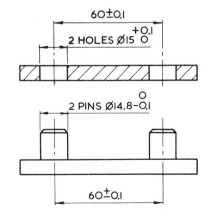

Fig. 33.30

allowed by the specified tolerances. The male component has centre distance and diameters of pins at maximum condition. The female component has centre distance and diameter of holes at minimum condition.

The tolerance diagram, fig. 33.31, shows that, when the pin diameters are at the least material condition, their centre distance may vary between 74.9 − 14.7 = 60.2, or 45.1 + 14.7 = 59.8. Now this increase in tolerance can be used to advantage, and can be obtained by applying the maximum material concept to the drawing detail.

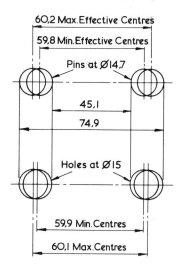

Fig. 33.31

Similarly, by applying the same principle to the female component, a corresponding advantage is obtainable. The lower part of fig. 33.31 shows the female component in its maximum material condition. Assembly with the male component will be possible if the dimension over the pins does not exceed 74.9 and the dimension between the pins is no less than 45.1.

Fig. 33.32 shows the method of dimensioning the female component with holes controlled by a positional tolerance, and modified by maximum material condition. This ensures assembly with the male component, whose pins are manufactured regardless of feature size.

When the maximum condition is applied to these features, any errors of form or position can be checked by using suitable gauges.

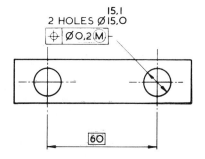

Fig. 33.32

Chapter 34

Co-ordinate and positional tolerances

The holes in the component shown in fig. 34.1 are positioned along one axis by toleranced co-ordinate dimensions. Since the hole centres between each pair of holes are toleranced separately, it follows that the limits are cumulative, e.g. between holes A and B they are ± 0.1, between holes A and C ± 0.2, and between holes A and D ± 0.3.

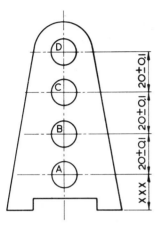

Fig. 34.1

Fig. 34.2 shows an alternative method of applying the co-ordinate method, from a common datum which is the centre line of hole A. Note that the distances between holes B and C could be either 20.2 or 19.8, i.e. 20 ± 0.2, depending on whether holes B and C are drilled at their high or low limits of distance with respect to hole A. Similarly, between holes B and D the distance could be 40.2 or 39.8, i.e. 40 ± 0.2.

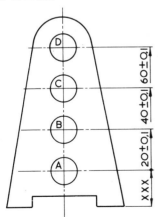

Fig. 34.2

No further accumulation of error results after the first pair of hole pitches if this method is used on a long chain of holes; nevertheless, if the acceptable practical error is only ± 0.1, then the drawing must have limits of ± 0.05 applied for the above reason.

A further example of toleranced co-ordinate dimensioning is shown in fig. 34.3, where two holes are positioned from datum edges.

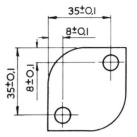

Fig. 34.3

This type of dimensioning assumes that the two datum edges are exactly at 90° to each other. Should this not be the case, then the distance between the co-ordinate tolerances zones shown in fig. 34.4 could vary considerably when the angle is greater or less than 90°.

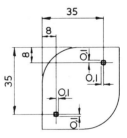

Fig. 34.4

True-position dimensioning

True-position dimensioning defines the exact location on a component of features such as holes, slots, keyways, etc., and also differentiates between 'ideal' and other toleranced dimensions. True-position dimensions are always shown 'boxed' on engineering drawings; they are never individually toleranced, and must always be accompanied by a positional or zone tolerance for the feature to which they are applied.

The positional tolerance is the permitted deviation of a feature from a true position.

The positional-tolerance zone defines the region which contains the extreme limits of position and can be, for example, rectangular, circular, cylindrical, etc.

Typical product requirement In the example shown in figs 34.5 and 34.6, the hole axis must lie within the cylindrical tolerance zone fixed by the true-position dimensions.

Some advantages of using this method are:
1 interpretation is easier, since true boxed dimensions fix the exact positions of details;
2 there are no cumulative tolerances;

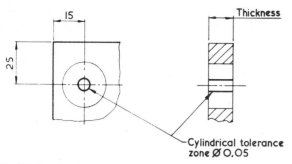

Fig. 34.5   Product requirement

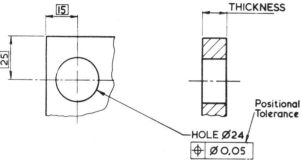

Fig. 34.6   Drawing instruction

3 it permits the use of functional gauges to match the mating part;
4 it can ensure interchangeability without resorting to small position tolerances, required by the co-ordinate tolerancing system;
5 the tolerancing of complicated components is simplified;
6 positional-tolerance zones can control squareness and parallelism.

The following examples show some typical cases where positional tolerances are applied to engineering drawings.

Case 1 (figs 34.7 and 34.8). The axes of the four fixing holes must be contained within cylindrical tolerance zones 0.03 diameter.

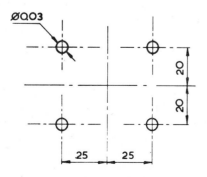

Fig. 34.7   Case 1: Product requirement

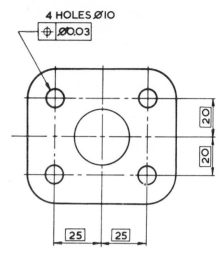

Fig. 34.8   Case 1: Drawing instruction

Case 2 (figs 34.9 and 34.10). The axes of the four fixing holes must be contained within rectangular tolerance zones 0.04 x 0.02.

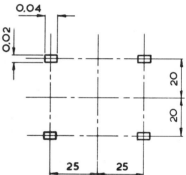

Fig. 34.9   Case 2: Product requirement

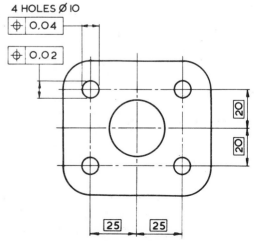

Fig. 34.10 Case 2: Drawing instruction

Case 3. Fig. 34.11 shows a component where the outside diameter at the upper end is required to be square and concentric within a combined tolerance zone with face A and diameter B as the primary and secondary datums.

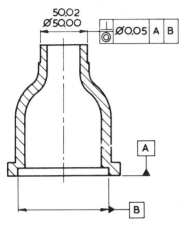

Fig. 34.11   Case 3

Case 4 In the component illustrated in fig. 34.12, the three dimensioned features are required to be perfectly square to the datum face A, and also truly concentric with each other in the maximum material condition.

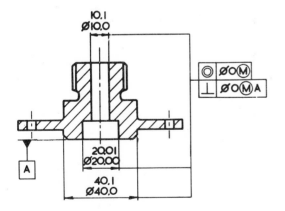

Fig. 34.12   Case 4

Case 5 (figs 34.13 and 34.14). The six bolt-holes on the flange in fig. 34.13 must have their centres positioned within six tolerance zones of ∅0.25 when the bolt holes are at their maximum material condition (i.e. minimum limit of size).

Note in fig. 34.14 that all the features in the group have the same positional tolerance in relation to each other. This method also limits in all directions the relative displacement of each of the features to each other.

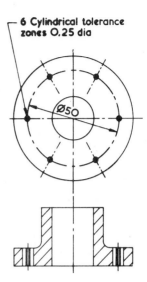

Fig. 34.13   Case 5: Product requirement

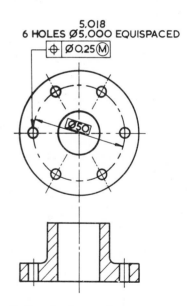

Fig. 34.14   Case 5: Drawing instruction

Case 6 (figs 34.15 and 34.16). The group of holes in fig. 34.15, dimensioned with a positional tolerance, is also required to be positioned with respect to the datum spigot and the face of the flange.

Note in fig. 34.16 that the four holes and the spigot are dimensioned at the maximum material condition. It follows that, if any hole is larger than 12.00, it will have the effect of increasing the positional tolerance for that hole. If the spigot is machined to less than 50.05, then the positional tolerance for the four holes as a group will also increase.

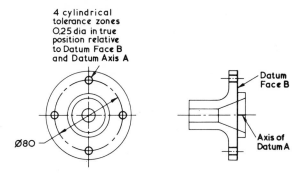

Fig. 34.15 Case 6: Product requirement

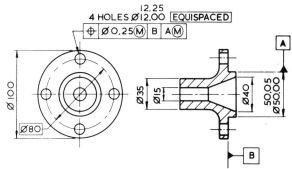

Fig. 34.16 Case 6: Drawing instruction

Case 7 Fig. 34.17 shows a drawing instruction where the group of equally spaced holes is required to be positioned relative to a concentric datum bore.

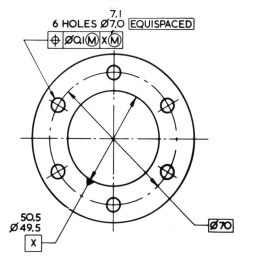

Fig. 34.17 Case 7

Case 8 In the example in fig. 34.18, control of the two $\emptyset10$ holes is required with respect to the datum hole X. The drawing instruction implies that the two $\emptyset10$ holes must be

in line, dictated by the boxed dimension, but subject to a positional tolerance of $\emptyset0.3$. In addition, their centres are subject to variation by virtue of the positional tolerance given to the $\emptyset16$ datum hole. Fig. 34.19 shows an extreme but acceptable condition for the position of the hole centres.

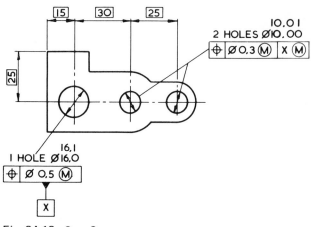

Fig. 34.18 Case 8

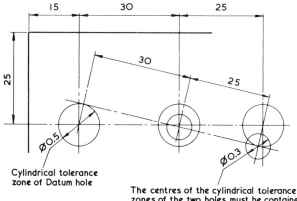

Fig. 34.19 Case 8: Extreme condition

Case 9 The example in fig. 34.20 shows three holes, of two different sizes, subject to the same positional tolerance. Since co-ordinate toleranced dimensions fix the position of the first hole, and true boxed dimensions fix the distance between holes, the extreme but acceptable condition shown in fig. 34.21 can arise.

Case 10 The situation in case 9 can be extended to a larger group of holes fixed by boxed dimensions, as fig. 34.22 illustrates. The extreme but acceptable condition shown in fig. 34.23 can arise.

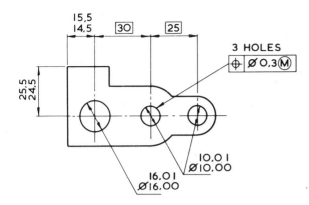

Fig. 34.20 Case 9

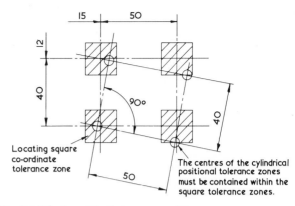

Fig. 34.23 Case 10: Extreme condition

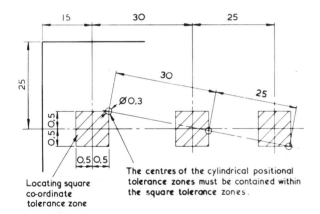

Fig. 34.21 Case 9: Extreme condition

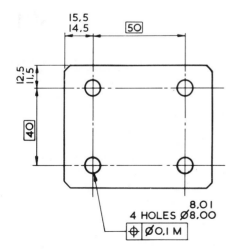

Fig. 34.22 Case 10

Chapter 35

Cams

A cam is generally a disc or a cylinder mounted on a rotating shaft, and it gives a special motion to a *follower*, by direct contact. The cam profile is determined by the required follower motion and the design of the type of follower.

The motions of cams can be considered to some extent as alternatives to motions obtained from linkages, but they are generally easier to design, and the resulting actions can be accurately predicted. If, for example, a follower is required to remain stationary, then this is achieved by a concentric circular arc on the cam. For a specified velocity or acceleration, the displacement of the follower can easily be calculated, but these motions are very difficult to arrange precisely with linkages.

Specialist cam-manufacturers computerise design data and, for a given requirement, would provide a read-out with cam dimensions for each degree, minute, and second of camshaft rotation.

When used in high-speed machinery, cams may require to be balanced, and this becomes easier to perform if the cam is basically as small as possible. A well-designed cam system will involve not only consideration of velocity and acceleration but also the effects of out-of-balance forces, and vibrations. Suitable materials must be selected to withstand wear and the effect of surface stresses.

Probably the most widely used cam is the *plate cam*, with its contour around the circumference. The line of action of the follower is usually either vertical or parallel to the camshaft, and fig. 35.1 shows several examples.

Examples are given later of a *cylindrical* or *drum cam*, where the cam groove is machined around the circumference, and also a *face cam*, where the cam groove is machined on a flat surface.

Cam followers

Various types of cam followers are shown in fig. 35.1.

Knife-edge followers are restricted to use with slow-moving mechanisms, due to their rapid rates of wear. Improved stability can be obtained from the roller follower, and increased surface area in contact with the cam can be obtained from the flat and mushroom types of follower. The roller follower is the most expensive type, but is ideally suited to high speeds and applications where heat and wear are factors.

Cam follower motions

1 Uniform velocity. This motion is used where the follower is required to rise or fall at a constant speed, and is often referred to as 'straight-line' motion. Part of a uniform-velocity cam graph is shown in fig. 35.2.

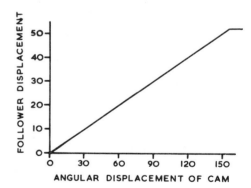

Fig. 35.2

Abrupt changes in velocity with high-speed cams result in large accelerations and cause the followers to jerk or chatter. To reduce the shock on the follower, the cam graph can be modified as indicated in fig. 35.3, by adding radii to remove the sharp corners. However, this action results in an increase in the average rate of rise or fall of the follower.

2 Uniform acceleration and retardation motion is shown in fig. 35.4. The graphs for both parts of the motion are parabolic. The construction for the parabola involves dividing the cam-displacement angle into a convenient number of parts, and the follower displacement into the

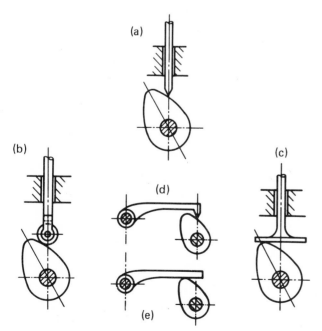

Fig. 35.1 Plate cams (a) Plate cam with knife-edge follower (b) Plate cam with roller follower (c) Plate cam with flat follower (d) Plate cam with oscillating knife-edge follower (e) Plate cam with oscillating flat follower

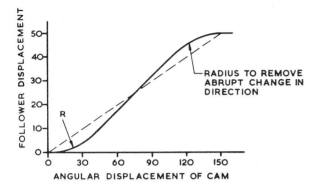

Fig. 35.3

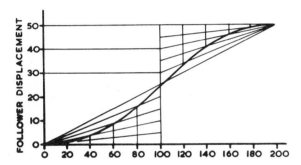

Fig. 35.4

same number of parts. Radial lines are drawn from the start position to each of the follower division lines, and the parabola is obtained by drawing a line through successive intersections. The uniform-retardation parabola is constructed in a similar manner, but in the reverse position.

3 Simple harmonic motion is shown in fig. 35.5, where the graph is a sine curve. The construction involves drawing a semi-circle and dividing it into the same number of parts as the cam-displacement angle. The diameter of the semi-circle is equal to the rise or fall of the follower. The graph passes through successive intersections as indicated.

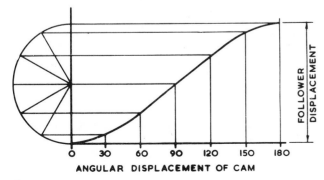

Fig. 35.5

The application of the various motions to different combinations of cams and followers is shown by the following practical examples.

Case 1 (fig. 35.6)

Cam specification:

Plate cam, rotating anticlockwise. Point follower. Least radius of cam, 30mm. Camshaft diameter, 20mm.

0–90°, follower rises 20mm with uniform velocity.

90–150°, follower rises 30mm with simple harmonic motion.

150–210°, dwell period.

210–270°, follower falls 20mm with uniform acceleration.

270–360°, follower falls 30mm with uniform retardation.

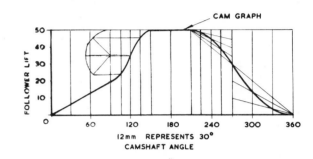

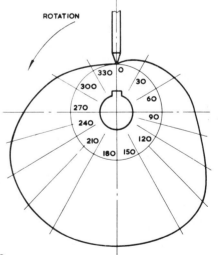

Fig. 35.6

1 Draw the graph as shown. Exact dimensions are used for the *Y* axis, where the follower lift is plotted. The *X* axis has been drawn to scale, where 12mm represents 30° of shaft rotation.

2 To plot the cam, draw a 20mm diameter circle to represent the bore for the camshaft, and another circle 30mm radius to represent the base circle, or the least radius of the cam, i.e. the nearest the follower approaches to the centre of rotation.

3 Draw radial lines 30° apart from the cam centre, and number them in the reverse direction to the cam rotation.

4 Plot the Y ordinates from the cam graph along each of the radial lines in turn, measuring from the base circle. Where rapid changes in direction occur, or where there is uncertainty regarding the position of the profile, more points can be plotted at 10° or 15° intervals.

5 Draw the best curve through the points to give the required cam profile.

Note. The user will require to know where the cam program commences, and the zero can be conveniently established on the same centre line as the shaft keyway. Alternatively, a timing hole can be drilled on the plate, or a mark may be engraved on the plate surface. In cases where the cam can be fitted back to front, the direction of rotation should also be clearly marked.

Case 2 (fig. 35.7)

Cam specification:

Plate cam, rotating anticlockwise. Flat follower. Least distance from follower to cam centre, 30mm. Camshaft diameter, 20mm.

0–120°, follower rises 30mm with uniform velocity (modified).

120–210°, dwell period.

210–360°, follower falls 30mm with uniform velocity (modified).

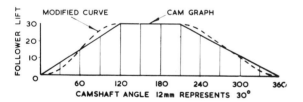

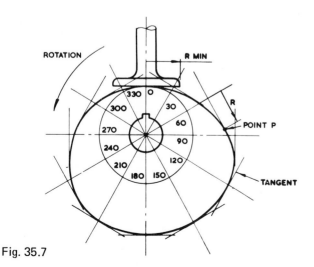

Fig. 35.7

1 Draw the cam graph as shown, and modify the curve to remove the sharp corners. Note that in practice the size of the radius frequently used here varies between one third and the full lift of the follower for the uniform-velocity part of the graph; the actual value depends on the rate of velocity and the speed of rotation. This type of motion is not desirable for high speeds.

2 Draw the base circle as before, 30mm radius, divide it into 30° intervals, and number them in the reverse order to the direction of rotation.

3 Plot the Y ordinates from the graph, radially from the base circle along each 30° interval line.

Draw a tangent at each of the plotted points, as shown, and draw the best curve to touch the tangents. The tangents represent the face of the flat follower in each position.

4 Check the point of contact between the curve and each tangent and its distance from the radial line. Mark the position of the widest point of contact. In the illustration given, point P appears to be the greatest distance, and hence the follower will require to be at least R in radius to keep in contact with the cam profile at this point. Note also that a flat follower can be used only where the cam profile is always convex.

Although the axis of the follower and the face are at 90° in this example, other angles are in common use.

Case 3 (fig. 35.8).

Cam specification:

Plate cam, rotating clockwise. 20mm diameter roller follower.

30mm diameter camshaft. Least radius of cam, 35mm.

0–180°, rise 64mm with simple harmonic motion.

180–240°, dwell period.

240–360°, fall 64mm with uniform velocity.

1 Draw the cam graph as shown.

2 Draw a circle (shown as RAD Q) equal to the least radius of the cam plus the radius of the roller, and divide it into 30° divisions. Mark the camshaft angles in the anticlockwise direction.

3 Along each radial line plot the Y ordinates from the graph, and at each point draw a 20mm circle to represent the roller.

4 Draw the best profile for the cam so that the cam touches the rollers tangentially, as shown.

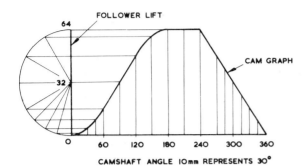

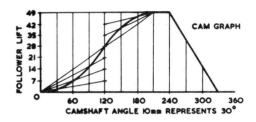

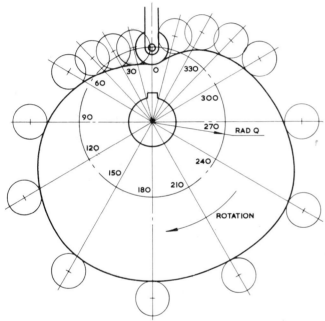

Fig. 35.8

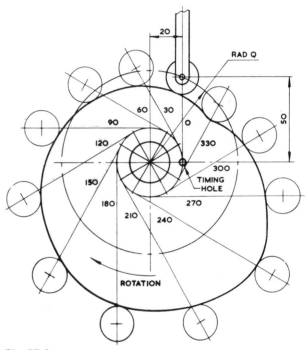

Fig. 35.9

Case 4 (fig. 35.9)

Cam specification:

Plate cam, rotating clockwise. 20mm diameter roller follower set 20mm to the right of the centre line for the camshaft. Least distance from the roller centre to the camshaft centre line, 50mm. 25mm diameter camshaft.

0–120°, follower rises 28mm with uniform acceleration.
120–210°, follower rises 21mm with uniform retardation.
210–240°, dwell period.
240–330°, follower falls 49mm with uniform velocity.
330–360°, dwell period.

1 Draw the cam graph as shown.
2 Draw a 20mm radius circle, and divide it into 30° divisions as shown.
3 Where the 30° lines touch the circumference of the 20mm circle, draw tangents at these points.

4 Draw a circle of radius *Q*, as shown, from the centre of the camshaft to the centre of the roller follower. This circle is the base circle.
5 From the base circle, mark lengths equal to the lengths of the *Y* ordinates from the graph, and at each position draw a 20mm diameter circle for the roller follower.
6 Draw the best profile for the cam so that the cam touches the rollers tangentially, as in the last example.

Case 5 (fig. 35.10)

Cam specification:

Face cam, rotating clockwise. 12mm diameter roller follower. Least radius of cam, 26mm. Camshaft diameter, 30mm.

0–180°, follower rises 30mm with simple harmonic motion.

180–240°, dwell period.

240–360°, follower falls 30mm with simple harmonic motion.

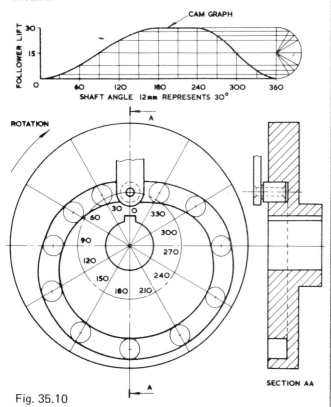

Fig. 35.10

1 Draw the cam graph, but note that for the first part of the motion the semi-circle is divided into six parts, and that for the second part it is divided into four parts.
2 Draw a base circle 32mm radius, and divide into 30° intervals.
3 From each of the base-circle points, plot the lengths of the *Y* ordinates. Draw a circle at each point for the roller follower.
4 Draw a curve on the inside and the outside, tangentially touching the rollers, for the cam track.

The drawing shows the completed cam together with a section through the vertical centre line.

Note that the follower runs in a track in this example. In the previous cases, a spring or some resistance is required to keep the follower in contact with the cam profile at all times.

Case 6 (fig. 35.11)

Cam specification:

Cylindrical cam, rotating anticlockwise, operating a roller follower 14mm diameter. Cam cylinder, 60mm diameter. Depth of groove, 7mm.

0–180°, follower moves 70mm to the right with simple harmonic motion.

180–360°, follower moves 70mm to the left with simple harmonic motion.

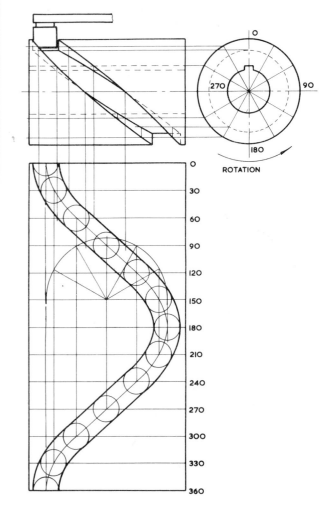

Fig. 35.11

1 Set out the cylinder blank and the end elevation as shown.
2 Divide the end elevation into 30° divisions.
3 Underneath the front elevation, draw a development of

the cylindrical cam surface, and on this surface draw the cam graph.

4 Using the cam graph as the centre line for each position of the roller, draw 14mm diameter circles as shown.

5 Draw the cam track with the sides tangential to the rollers.

6 Plot the track on the surface of the cylinder by projecting the sides of the track in the plan view up to the front elevation. Note that the projection lines for this operation do not come from the circles in the plan, except at each end of the track.

7 The dotted line in the end elevation indicates the depth of the track.

8 Plot the depth of the track in the front elevation from the end elevation, as shown. Join the plotted points to complete the front elevation.

Note that, although the roller shown is parallel, tapered rollers are often used, since points on the outside of the cylinder travel at a greater linear speed than points at the bottom of the groove, and a parallel roller follower tends to skid.

Dimensioning cams

Fig. 35.12 shows a cam in contact with a roller follower; note that the point of contact between the cam and the roller is at A, on a line which joins the centres of the two arcs. To dimension a cam, the easiest method of presenting the data is in tabular form which relates the angular displacement ϕ of the cam with the radial displacement R of the follower. The cam could then be cut on a milling machine using these point settings. For accurate master cams, these settings may be required at half- or one-degree intervals.

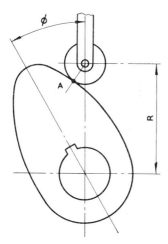

Fig. 35.12

Chapter 36

Gears

Spur gears

The characteristic feature of spur gears is that their axes are parallel. The gear teeth are positioned around the circumference of the pitch circles which are equivalent to the circumferences of the friction rollers in fig. 36.1 below.

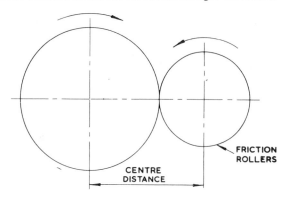

Fig. 36.1

The teeth are of *involute* form, the involute being described as the locus traced by a point on a taut string as it unwinds from a circle, known as the base circle. For an involute rack, the base-circle radius is of infinite length, and the tooth flank is therefore straight.

The construction for the involute profile is shown in fig. 36.2. The application of this profile to an engineering drawing of a gear tooth can be rather a tedious exercise, and approximate methods are used, as described later.

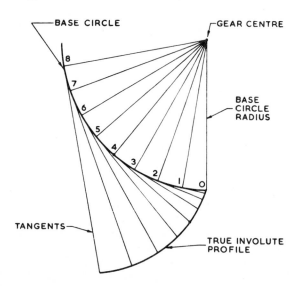

Fig. 36.2   Involute construction. The distance along the tangent from each point is equal to the distance around the circumference from point O.

Spur-gear terms

The *gear ratio* is the ratio of the number of teeth in the gear to the number of teeth in the pinion, the pinion being the smaller of the two gears in mesh.

The *pitch-circle diameters* of a pair of gears are the diameters of cylinders co-axial with the gears which will roll together without slip. The pitch circles are imaginary friction discs, and they touch at the *pitch point*.

The *base circle* is the circle from which the involute is generated.

The *root diameter* is the diameter at the base of the tooth.

The *centre distance* is the sum of the pitch-circle radii of the two gears in mesh.

The *addendum* is the radial height of the tooth from the pitch circle to the tooth tip.

The *dedendum* is the radial depth of the tooth from the pitch circle to the root of the tooth.

The *clearance* is the algebraic difference between the addendum and the dedendum.

The *whole depth* of the tooth is the sum of the addendum and the dedendum.

The *circular pitch* is the distance from a point on one tooth to the corresponding point on the next tooth, measured round the pitch-circle circumference.

The *tooth width* is the length of arc from one side of the tooth to the other, measured round the pitch-circle circumference.

The *module* is the pitch-circle diameter divided by the number of teeth.

The *diametral pitch* is the reciprocal of the module, i.e. the number of teeth divided by the pitch-circle diameter.

The *line of action* is the common tangent to the base circles, and the *path of contact* is that part of the line of action where contact takes place between the teeth.

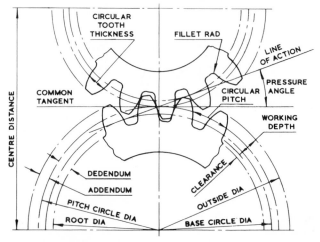

Fig. 36.3   Spur-gear terms

142

The *pressure angle* is the angle formed between the common tangent and the line of action.

The *fillet* is the rounded portion at the bottom of the tooth space.

The various terms are illustrated in fig. 36.3.

Involute gear teeth proportions and relationships

Module = $\dfrac{\text{pitch-circle diameter, PCD}}{\text{number of teeth, } T}$

Circular pitch = π x module

Tooth thickness = $\dfrac{\text{circular pitch}}{2}$

Addendum = module
Clearance = 0.25 x module
Dedendum = addendum + clearance

Involute gears having the same pressure angle and module will mesh together. The British Standard recommendation for pressure angle is 20°.

The conventional representation of gears shown in fig. 36.4 is limited to drawing the pitch circles and outside diameters in each case. In the sectional end elevation, a section through a tooth space is taken as indicated. This convention is common practice with other types of gears and worms.

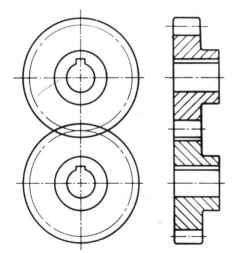

Fig. 36.4   Gear conventions

Typical example using Professor Unwin's approximate construction

Gear data:
Pressure angle, 20°. Module, 12mm: Number of teeth, 25.

Gear calculations:
Pitch-circle diameter = module x no. of teeth
= 12 x 25 = 300mm
Addendum = module = 12mm
Clearance   = 0.25 x module
= 0.25 x 12 = 3mm
Dedendum = addendum + clearance = 12 + 3 = 15mm
Circular pitch = π x module = π x 12 = 37.68mm
Tooth thickness = ½ x circular pitch = 18.84mm

Stage 1 (fig. 36.5)
a) Draw the pitch circle and the common tangent.
b) Mark out the pressure angle and the normal to the line of action.
c) Draw the base circle. Note that the length of the normal is the base-circle radius.

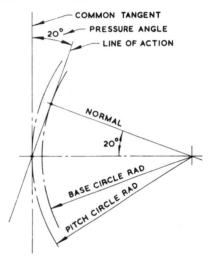

Fig. 36.5   Unwin's construction—stage 1

Stage 2 (fig. 36.6)
a) Draw the addendum and dedendum circles. Both addendum and dedendum are measured radially from the pitch circle.
b) Mark out point A on the addendum circle and point B on the dedendum circles. Divide AB into three parts so that CB = 2AC.
c) Draw the tangent CD to the base circle. D is the point of tangency. Divide CD into four parts so that CE = 3DE.
d) Draw a circle with centre O and radius OE. Use this circle for centres of arcs of radius EC for the flanks of the teeth after marking out the tooth widths and spaces around the pitch-circle circumference.

Note that it may be more convenient to establish the length of the radius CE by drawing this part of the construction further round the pitch circle, in a vacant space, if the flank of one tooth, i.e. the pitch point, is required to lie on the line AO.

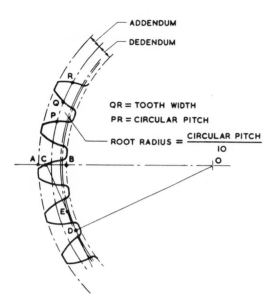

Fig. 36.6   Unwin's construction—stage 2

The construction is repeated in fig. 36.7 to illustrate an application with a rack and pinion. The pitch line of the rack touches the pitch circle of the gear, and the values of the addendum and dedendum for the rack are the same as those for the meshing gear.

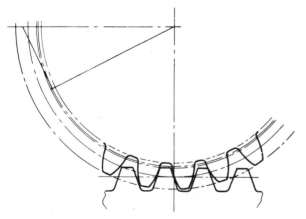

Fig. 36.7   Unwin's construction applied to a rack and pinion

If it is required to use the involute profile instead of the approximate construction, then the involute must be constructed from the base circle as shown in fig. 36.2. Complete stage 1 and stage 2(a) as already described, and mark off the tooth widths around the pitch circle, commencing at the pitch point. Take a tracing of the involute in soft pencil on transparent tracing paper, together with part of the base circle in order to get the profile correctly oriented on the required drawing. Using a French curve, mark the profile in pencil on either side of the tracing paper, so that, whichever side is used, a pencil impression can be obtained. With care, the profile can now be traced onto the required layout, lining up the base circle and ensuring that the profile of the tooth flank passes through the tooth widths previously marked out on the pitch circle. The flanks of each tooth will be traced from either side of the drawing paper. Finish off each tooth by adding the root radius.

Helical gears

Helical gears have been developed from spur gears, and their teeth lie at an angle to the axis of the shaft. Contact between the teeth in mesh acts along the diagonal face flanks in a progressive manner; at no time is the full length of any one tooth completely engaged. Before contact ceases between one pair of teeth, engagement commences between the following pair. Engagement is therefore continuous, and this fact results in a reduction of the shock which occurs when straight teeth operate under heavy loads. Helical teeth give a smooth, quiet action under heavy loads; backlash is considerably reduced; and, due to the increase in length of the tooth, for the same thickness of gear wheel, the tooth strength is improved.

Fig. 36.8 illustrates the lead and helix angle applied to a helical gear. For single helical gears, the helix angle is generally $12°-20°$.

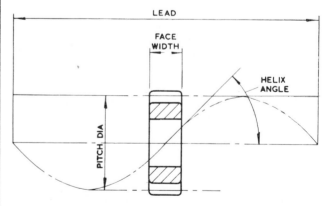

Fig. 36.8   Lead and helix angle for a helical gear

Since the teeth lie at an angle, a side or end thrust occurs when two gears are engaged, and this tends to separate the gears. Fig. 36.9 shows two gears on parallel shafts and the position of suitable thrust bearings. Note that the position of the thrust bearings varies with the direction of shaft rotation and the 'hand' of the helix.

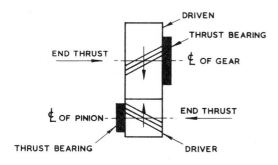

Fig. 36.9

In order to eliminate the serious effect of end thrust, pairs of gears may be arranged as shown in fig. 36.10, where a double helical gear utilises a left- and a right-hand helix. Instead of using two gears, the two helices may be cut on the same gear blank.

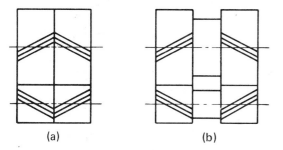

| (a) | (b) |

Fig. 36.10 Double helical gears. (a) On same wheel (b) On separate wheels

Where shafts lie parallel to each other, the helix angle is generally 15°–30°. Note that a right-hand helix engages with a left-hand helix, and the hand of the helix must be correctly stated on the drawing. On both gears the helix angle will be the same.

For shafts lying at 90° to each other, both gears will have the same hand of helix, fig. 36.11.

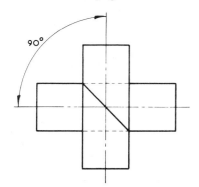

Fig. 36.11

Helical gears can be used for shafts which lie at an angle less than 90°, but the hand of helix should be checked with a specialist gear manufacturer. The hand of a helix depends on the helix angle used and the shaft angles required.

Bevel gears

If the action of spur and helical gears can be related to that of rolling cylinders, then the action of bevel gears can be compared to a friction cone drive. Bevel gears are used to connect shafts which lie in the same plane and whose axes intersect. The size of the tooth decreases as it passes from the back edge towards the apex of the pitch cone, hence the cross-section varies along the whole length of the tooth. When viewed on the curved surface which forms part of the back cone, the teeth normally have the same profiles as spur gears. The addendum and dedendum have the same proportions as a spur gear, being measured radially from the pitch circle, parallel to the pitch-cone generator.

Data relating to bevel gear teeth is shown in fig. 36.12. Note that the crown wheel is a bevel gear where the pitch angle is 90°. Mitre gears are bevel gears where the pitch-cone angle is 45°.

The teeth on a bevel gear may be produced in several different ways, e.g. straight, spiral, helical, or spiraloid The advantages of spiral bevels over straight bevels lies in quieter running at high speed and greater load-carrying capacity.

The angle between the shafts is generally a right angle, but may be greater or less than 90°, as shown in fig. 36.13.

Bevel gearing is used extensively in the automotive industry for the differential gearing connecting the drive shaft to the back axle of motor vehicles.

Bevel-gear terms and definitions
The following are additions to those terms used for spur gears.

The pitch angle is the angle between the axis of the gear and the cone generating line.

The root-cone angle is the angle between the gear axis and the root generating line.

The face angle is the angle between the tips of the teeth and the axis of the gear.

The addendum angle is the angle between the top of the teeth and the pitch-cone generator.

The dedendum angle is the angle between the bottom of the teeth and the pitch-cone generator.

The outside diameter is the diameter measured over the tips of the teeth.

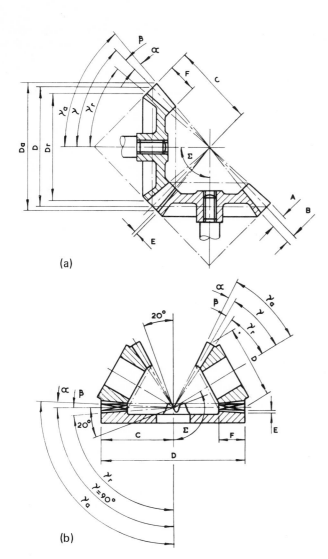

(a)

(b)

Fig. 36.12 Bevel-gear terms (a) Bevel gears (b) Crown
            wheel and pinions
            A = addendum
            B = dedendum
            C = cone distance
            D = pitch diameter
            Da = outside diameter
            Dr = root diameter
            E = bottom clearance
            F = face width
            α (alpha) = addendum angle
            β (beta) = dedendum angle
            γ (gamma) = pitch angle
            γa = back cone angle
            γr = root angle
            Σ (sigma) = shaft angle

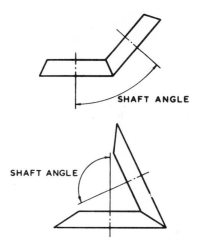

Fig. 36.13 Bevel-gear cones

The following figures show the various stages in drawing
bevel gears. The approximate construction for the profile of
the teeth has been described in the section relating to spur
gears.

Gear data: 15 teeth, 20° pitch-cone angle, 100mm
pitch-circle diameter, 20° pressure angle.

Stage 1 Set out the cone as shown in fig. 36.14.

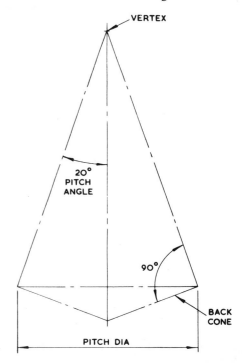

Fig. 36.14 Stage 1

Stage 2 Set out the addendum and dedendum. Project part of the auxiliary view to draw the teeth (fig. 36.15).

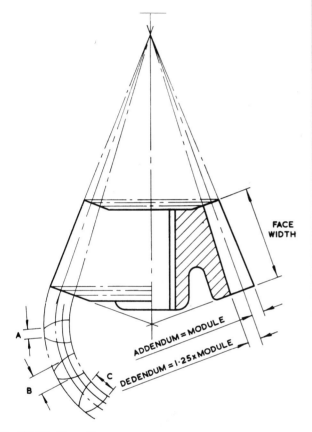

FACE WIDTH

ADDENDUM = MODULE

DEDENDUM = 1·25 × MODULE

A

B

C

Fig. 36.15 Stage 2

Stage 3 Project widths A. B, and C on the outside, pitch, and root diameters, in plan view.
Complete the front elevation (fig. 36.16).

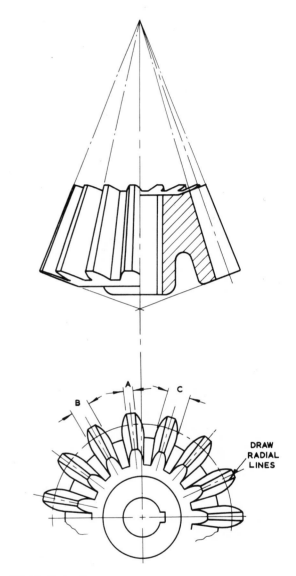

A

B

C

DRAW RADIAL LINES

Fig. 36.16 Stage 3

Worm gearing

Worm gearing is used to transmit power between two non-intersecting shafts whose axes lie at right angles to each other. The drive of a worm gear is basically a screw, or worm, which may have a single- or multi-start thread, and this engages with the wheel. A single-start worm in one revolution will rotate the worm wheel one tooth and space, i.e. one pitch. The velocity ratio is high; for example, a 40 tooth wheel with a single-start worm will have a velocity ratio of 40, and in mesh with a two-start thread the same wheel would give a velocity ratio of 20.

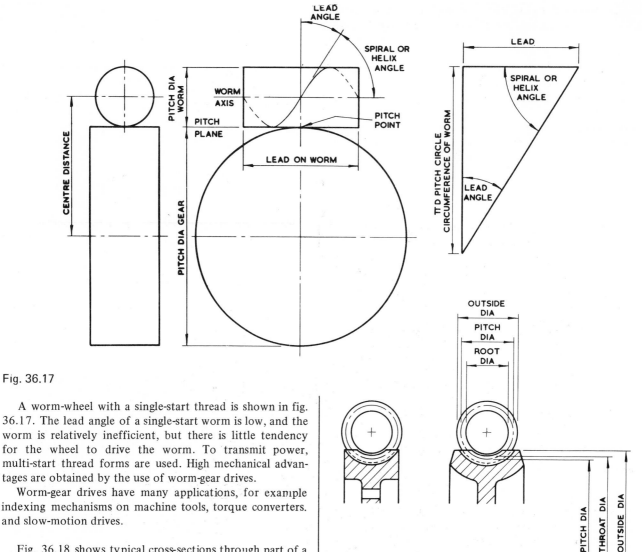

Fig. 36.17

A worm-wheel with a single-start thread is shown in fig. 36.17. The lead angle of a single-start worm is low, and the worm is relatively inefficient, but there is little tendency for the wheel to drive the worm. To transmit power, multi-start thread forms are used. High mechanical advantages are obtained by the use of worm-gear drives.

Worm-gear drives have many applications, for example indexing mechanisms on machine tools, torque converters, and slow-motion drives.

Fig. 36.18 shows typical cross-sections through part of a worm and wheel. Note the contour of the wheel, which is designed to give greater contact with the worm.

Fig. 36.18   Worm-gearing terms applied to a worm and part of a worm-wheel

Chapter 37

Ball and roller bearings

Ball and roller bearings are widely applied in all branches of industry, both as original installations and as replacements for plain bearings. In this type of bearing, rolling friction is substituted for sliding friction.

Sliding friction is the resistance to motion incurred when one surface slides over another surface; the surfaces may not necessarily be similar. A magnified cross-section would reveal 'hills and valleys', and these tend to interlock when sliding takes place, so that forces of considerable magnitude are required for motion. The separation of these surfaces by a film of grease or oil is a fundamental aim of lubrication. Rolling friction, or the resistance to rolling, normally constitutes a relatively small impediment to motion, and is preferred. When machinery using plain bearings is started from rest, the initial resistance to motion is considerably in excess of the resistance which is offered to maintaining motion, after the bearings have built up their fluid film of lubricant to keep the rubbing surfaces apart. Where ball and roller bearings are used, the resistance to motion from rest is only very little more than the resistance to continued motion, and in both cases is much less than that encountered in comparable plain bearings. This advantage has been responsible for the widening sphere of application of roller bearings, particularly in machinery which is subject to frequent starting, stopping, or reversing.

The British standard convention for a bearing is shown in fig. 37.1. Note that for each type of bearing the convention is the same.

Bearing terminology

Fig. 37.2 shows various parts of bearings, to illustrate the terminology in common use.

Selection of bearings

The following typical arrangements of bearings and the diagrams of bearing terminology overleaf are reproduced by kind permission of Ransome, Hoffmann, Pollard Ltd, of Chelmsford, Essex. For any particular use, the successful performance of a bearing in service will depend on all aspects of the application being considered at the design stage. The manufacturer's catalogue should be consulted for a complete specification of each bearing required.

Bearing-mounting arrangements

a) the bearing types used must be suitable for the load and operating conditions,
b) the fits demanded by the mounting requirements must be compatible with those necessary for the load and rotation conditions,
c) the shaft must be located axially in both directions to the limit required by the application,
d) the mounting arrangement must be able to accom-

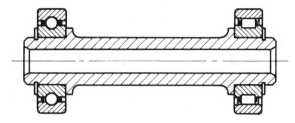

Ball and roller bearings

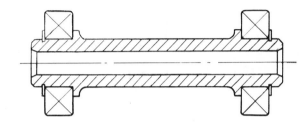

BS conventions

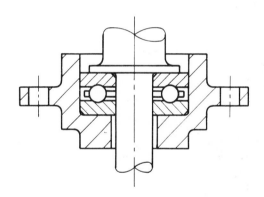

Single-direction thrust ball bearing

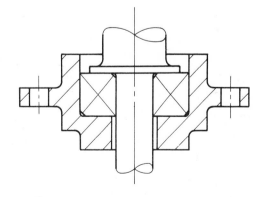

BS convention

Fig. 37.1　BS conventional representation of bearings

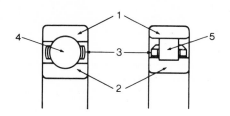

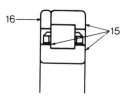

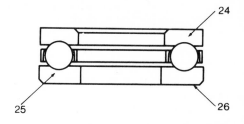

1   Outer ring
2   Inner ring
3   Cage
4   Ball ⎫ Rolling
5   Cylindrical roller ⎭ elements

15   Ribs
16   Loose rib

24   Large bore washer
25   Small bore washer
26   Flat seat

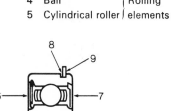

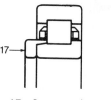

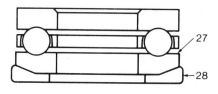

6   Seal
7   Shield
8   Snap ring groove
9   Locating snap ring

17   Separate thrust collar

27   Aligning washer
28   Aligning seat washer

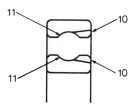

10   Filling slots
11   Raceways

18   Spherical end ⎫
19   Ball end ⎬ Needle
20   Conical end ⎪ rollers
21   Straight trunnion end ⎭

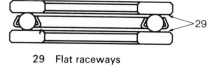

29   Flat raceways

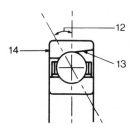

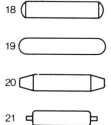

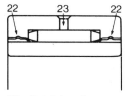

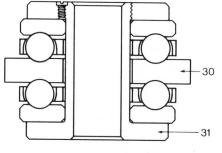

12   Contact angle
13   Counterbore
14   Thrust face

22   Retaining ring
23   Oil hole

30   Large outside diameter
      centre washer
31   Sleeve

Fig. 37.2   Bearing terminology

modate any thermal differences between the shaft and the housing.

The typical mountings are shown here, and can be adapted to suit the majority of engineering applications.

Bearing-mounting arrangements: horizontal shafts

Two single-row radial ball bearings Used when the shaft rotates in relation to the load line.

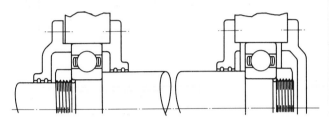

Loading Both bearings are suitable for dealing with radial loads, and the locating bearing takes axial loads in either direction.

Seating fits Inner rings—interference fit to prevent creep; outer rings—sliding fit to allow the non-locating bearing outer ring to take up the correct axial position without preload.

Axial location The locating bearing positions the shaft axially in both directions with some end movement.

Single-row radial ball bearing—snap-ring type The mounting of the location bearing is often simplified by using a snap-ring type as shown in the diagram below.

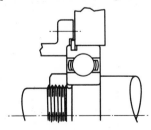

Two single-row radial ball bearings—diagonal location Used where the shaft rotates in relation to the load line. The diagram also illustrates the use of single shielded bearings, which simplify protection and grease retention.

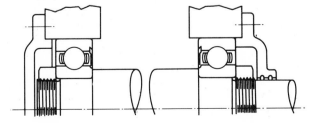

Loading Both bearings deal with radial loads, and each bearing takes axial load in one direction only.

Seating fits Inner rings—interference fit to prevent creep; outer rings—sliding fit to allow the bearings to assume the correct axial position without preload.

Axial location Each bearing locates the shaft axially in one direction. Sufficient axial clearance must be provided at the end-cover abutment face to allow for positional tolerances and any relative thermal expansion of the shaft and housing.

Two single-row radial ball bearings Used where the housing rotates in relation to the load line.

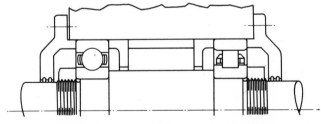

Loading Both bearings deal with the radial loads, and the locating bearing takes axial loads in either direction.

Seating fits Inner rings—these are made a sliding fit to allow the non-locating bearing to take up its correct axial position; outer rings—interference fit to prevent creep.

Axial location The locating bearing positions the housing axially in both directions with some end movement.

One single-row radial ball bearing with one cylindrical roller bearing Used for any combination of load and rotation.

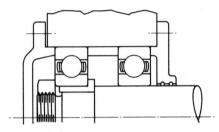

Loading The ball bearing deals with both radial and axial loading. The roller bearing is at the end of the shaft where the radial load is heaviest.

Seating fits The rings that rotate in relation to the load line must be made an interference fit to avoid creep. The other rings may be a sliding fit to assist assembly, but may be made an interference fit if necessary.

Axial location The ball bearing locates the shaft axially in both directions, with some end movement. The non-ribbed outer ring of the roller bearing allows for positional tolerances and relative axial expansion without any danger or preload.

Two cylindrical roller bearings with single-rib outer rings
Used where the load line is constant in direction, or where
the load is rotating.

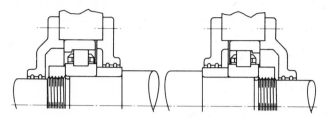

Loading Both bearings deal with radial loading, and the
bearing ribs may be used to support light and intermittent
axial loading, provided the rib-face/roller-end contacts are
effectively lubricated.
Seating fits Inner rings—interference fit to prevent creep;
outer rings—rings that rotate in relation to the load line
must be made an interference fit to avoid creep. The outer
rings may be a transition fit if clamped endways and if
there is no rotating load involved.
Axial location Each bearing locates the shaft axially in one
direction. Clearance between the roller ends and rib faces
must be provided to allow for positional tolerances and
relative axial expansion of shaft to housing.

**Two single-row angular contact ball bearings, face-to-face
mounting** Used for rotating shaft applications where the
load line is constant in direction.

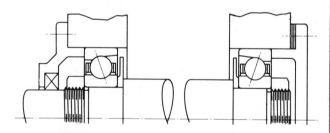

Loading Each bearing deals with radial loads, and axial
loads in one direction.
Seating fits Inner rings—interference fit to prevent creep;
outer rings—must be a sliding fit to permit correct axial
adjustment.
Axial location Each bearing locates the shaft axially in one
direction. The bearings are adjusted one against the other
(in this case by means of shims) to obtain correct ball
tracking and the required end movement. Care must be
taken to avoid excessive preload either initially or under
running conditions. Allowance must be made for relative
axial expansion of the shaft to the housing.

**Two single-row angular contact ball bearings, back-to-back
mounting** Used for rotating outer ring applications where
the load line is constant in direction.

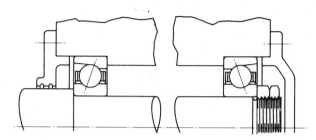

Loading Each bearing deals with radial loads, and axial
loads in one direction.
Seating fits Inner rings—these are made a sliding fit to allow
for correct axial adjustment (carried out in this case by a
screwed nut which must be finally locked in position);
outer rings—interference fit to prevent creep.
Axial location Each bearing locates the shaft axially in one
direction. The bearings are adjusted one against the other to
obtain correct ball tracking and the required end movement. Care must be taken during adjustment to avoid
excessive preload.

**One matched pair of angular contact ball bearings with one
cylindrical roller bearing** Used where close control of end
movement is necessary. Suitable for any conditions of
radial-load direction and rotation. For high speeds, if grease
lubrication is used, insert equal-length distance pieces
between the rings of the angular contact bearings.

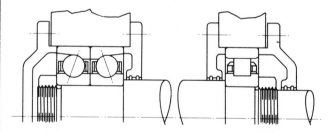

Loading The angular contact bearings support both radial
and axial loads. The roller bearing deals with radial load
only, and is usually positioned at the end of the shaft where
the radial load is heavier.
Seating fits The rings that rotate relative to the load line
must be an interference fit to prevent creep. The other rings
may be a sliding fit to assist assembly, but may be made an
interference fit if necessary.
Axial location The pre-adjusted angular contact bearings
locate the shaft in both directions. Consult the manufacturer if both rings are interference fits.

Two cylindrical roller bearings with one Duplex ball bearing (location pattern) Used for heavy loads and any conditions of direction and rotation.

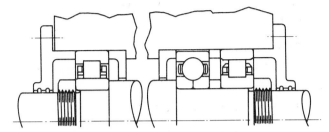

Loading The roller bearings take radial loads, the 'LOC' pattern Duplex bearing axial loads only. Sufficient load must be applied to the Duplex bearing under all conditions of rotation to ensure correct running.
Seating fits The roller-bearing rings that rotate relative to the radial load line must be made an interference fit to prevent creep. The other rings may be made a sliding or transition fit, or an interference fit if necessary. The Duplex bearing inner rings may be made a sliding fit on the shaft, although it may be more convenient to use the same fit as for the adjacent roller bearing.
Axial location In both directions with small end movement.

Spring loading
Spring loading is used to eliminate free play in the spindle. This is desirable for applications where quiet running or high rotational accuracy is required. Spring loading is often used to minimise the amount of rolling-element and raceway damage that occurs in stationary bearings subject to vibration.

Spring-box arrangement Used where the highest degree of rotational accuracy is required. The spring box should be as long as practicable to give accurate squareness control to the bearing face, and it should be made a sliding fit in the housing bore.

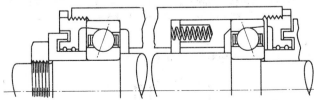

Loading The main axial load should be in the same direction as the spring load. Reverse thrust must always be less than the spring load, to maintain stability.
Seating fits Shaft—interference fit to prevent creep; housing—sliding fit to allow the spring loading to be correctly applied.

Axial location Each bearing locates the shaft axially in one direction with no free end movement.

Coil spring Used to apply a light preload to the bearings. It is not recommended for high-accuracy applications, as uneven spring pressure may cause out-of-squareness of the bearing outer ring.

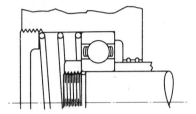

Preloading washers Preloading washers are often more convenient to use than coil-spring devices, and they take up less axial space. Various types are available. These washers should apply load evenly around the bearing outer ring, to avoid out-of-squareness. If high accuracy is required, a sliding-fit distance piece should be interposed between the washer and the bearing.

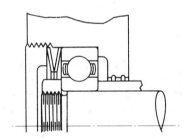

Bearing-mounting arrangements: vertical shafts
The principles already outlined apply to vertical shaft arrangements, but two further factors should be considered:
1 retention of lubricant.
2 shaft stability—for maximum stability the position of the locating bearing should be above the centre of gravity of the shaft.

One matched pair of single-row angular contact ball bearings with one cylindrical roller bearing. In this case a stationary baffle plate is used between the angular contact bearings to minimise the danger of grease slumping.
The baffle plate must be made to the exact width of the inner distance piece, to maintain the initial bearing adjustment. Separate lubrication points are provided for each bearing.

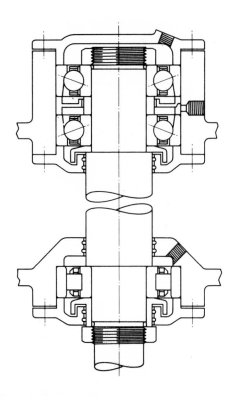

Selection of closure type

It is most important that the lubricant and the housing are kept clean and uncontaminated. Closures must prevent dirt, dust, and abrasive or corrosive media from entering the bearing housing. Bearings may also need protection from debris, such as metallic particles from adjacent gears.

Clearance-type closures These are effective for a wide range of conditions, and will retain both grease and oil, provided the oil level does not rise above the cover bore and there is no excessive splashing that would cause leakage. It is recommended that annular grooves are machined in the cover bore, and that these grooves are filled with a stiff grease to improve the effectiveness of the closure. The recommended clearance between the rotating and stationary parts under normal conditions is given below. Allowance must be made for shaft positional tolerances and thermal effects.

Shaft size		Radial clearance	Axial clearance
over	*incl.*	*(mm)*	*(mm)*
	25mm	0.12/0.25	0.25/0.5
25mm	100mm	0.25/0.38	0.5/0.75
100mm	—	0.5/0.64	1.0/1.25

Rubbing-seal closures These are effective in providing a contact seal on the rotating shaft, but the rubbing speed is limited. Shaft surface finish between 0.25 to 0.5 micrometres is normally recommended, and the seals should be properly lubricated to maintain their efficiency.

Typical closures

The following are examples of typical closures

Description
Annular grooved. Spigoted end cover

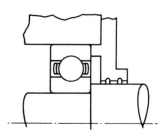

Application
Suitable for general engineering applications where normal indoor environment does not involve excessively dirty or dusty conditions.

Description
Single labyrinth closure

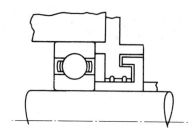

Application
Dirty or dusty conditions. A double labyrinth may be necessary if extremely dusty conditions are involved.

Description
Thrower-type labyrinth closure

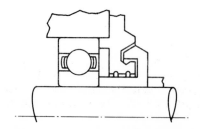

Application
Dirty and wet conditions. The rotating thrower and gutter machined in the end cover keep out dripping water.

Description
Thrower-type labyrinth closure

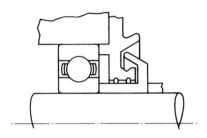

Application
Dirty and wet conditions. The rotating thrower and gutter in the end cover prevent splashing water from reaching the bearing.

Description
Thrower-type labyrinth closure with additional pressed metal cover and flinger.

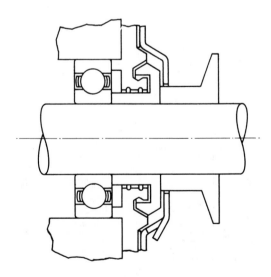

Application
Dirty and wet conditions where protection is required against water under pressure. The saucer-shaped flinger will deal with any water that works its way along the shaft, and any water that gains entry past the pressed metal cover is drained effectively through a hole at the base of this cover.

Description
Internal-type labyrinth and thrower.

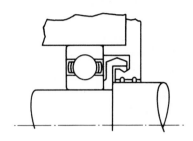

Application
Normal indoor environment where retention of splashing oil is required. The oil level should not rise above the cover bore under any conditions.

Description
Internal-type chip shield

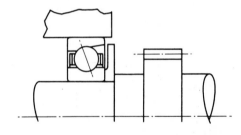

Application
The rotating chip shield is designed to protect the bearing against the metallic particles released by the adjacent gear whilst allowing oil splashes to reach the working surfaces.

Description
Blind end cover

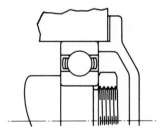

Application
Most environments where the housing must be sealed at the end of the shaft. The cover can be made completely oil-tight by inserting a gasket between the housing and cover faces.

Description
Felt seal

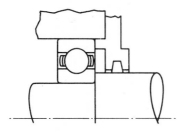

Application
Normal engineering applications where the rubbing speed does not exceed 7·65 m/s. Suitable for self-aligning bearings where the end-cover clearance required to accommodate the misalignment would be unacceptably large.

Description
Radial-lip seal. Usually designed to seal primarily in one direction, i.e. against ingress of foreign matter as shown, or against leakage of oil (seal mounted the other way round).

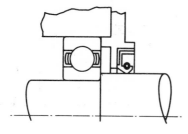

Application
Dirty and wet conditions. Will retain oil in the bearing housing. The recommendations given by the seal manufacturer for mounting and shaft requirements should be followed.

Description
Thrower-type labyrinth closure for vertical shaft.

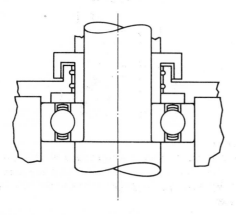

Application
Dirty and wet conditions. Suitable for use at the top of a vertical shaft.

Description
Labyrinth-type closure for a vertical shaft. With grease lubrication, the space above the bearing must be restricted to one quarter of the bearing width, and usually a no. 3 consistency grease is used to prevent grease slumping.

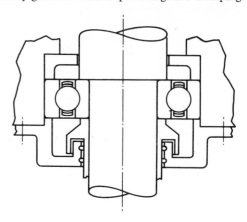

Application
Will retain lubricant at the bottom of a vertical shaft. If oil is the lubricant, the housing must not become flooded or leakage will occur.

Description
Recessed end cover

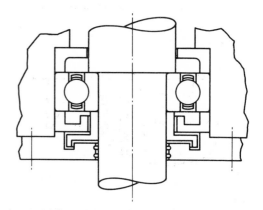

Application
Will retain grease at the bottom of a vertical shaft. More suited to the higher-speed applications, due to the omission of the internal thrower.

Description
Pressed metal labyrinth

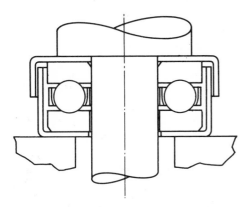

Application
Dirty and wet conditions for thrust ball bearings on a vertical shaft. Grease lubrication is employed, and if only slow and occasional rotation is involved the bearing may be completely filled with grease to give added protection.

Description
Fluid seal

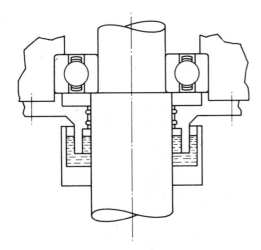

Application
Will prevent corrosive vapours from entering the housing of a vertical-shaft application, providing the speed is not so high as to cause excessive turbulence of the fluid.

Description
Shielded bearing with pressed metal shields fitted in the outer ring recesses, and running clearance on the inner ring. Single-shielded types also available.

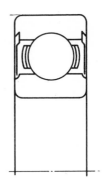

Application
Normal indoor environment where there are no excessively dirty or dusty conditons. The shields are not designed for removal, and the bearing is 'sealed for life'.

Description
Sealed bearing with bonded synthetic rubber seals fitted in the outer-ring recesses, and contact sealing on the inner-ring recesses. Single-sealed types also available.

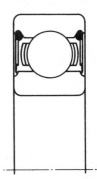

Application
Dirty conditions. The seals are not effective in retaining oil, due to the end movement in the bearing. See bearing tables for speed limitations. These bearings are 'sealed for life'.

Bearing lubrication
Oil bath lubrication
Suitable for most horizontal-shaft applications involving normal loads and speeds.

The optimum oil level is at the centre line of the lowest rolling element in the bearing. The surface area and volume of oil should be sufficiently large to maintain an adequate depth for the cage and rolling elements to dip into when running. A greater depth of oil could give rise to excessive oil churning and high temperatures—particularly if speeds are high.

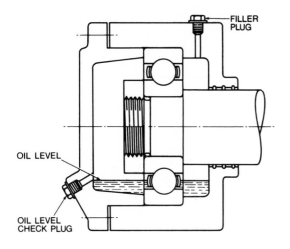

FILLER PLUG

OIL LEVEL

OIL LEVEL CHECK PLUG

Oil drip feed

This method is suitable for most horizontal and vertical shaft applications. A metering device is used to supply oil to the bearing and this can be regulated to suit the operating conditions. Attention must be given to drainage, to prevent over-filling the bearing housing, and it is usual on horizontal mountings to arrange the drain hole, or a weir, so that an oil bath is maintained at the correct level to provide lubrication on starting up.

Provision must be made to recirculate the oil or to allow it to run to waste.

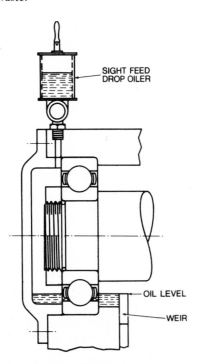

SIGHT FEED DROP OILER

OIL LEVEL

WEIR

Oil splash lubrication

Suitable for bearings in a gear-box when the oil used to lubricate the gears is adequately distributed to lubricate the bearings. The oil is either splashed directly onto the bearings, or collected in galleries from which it is fed through the bearings.

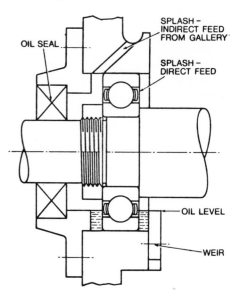

OIL SEAL

SPLASH – INDIRECT FEED FROM GALLERY

SPLASH – DIRECT FEED

OIL LEVEL

WEIR

Oil wick lubrication

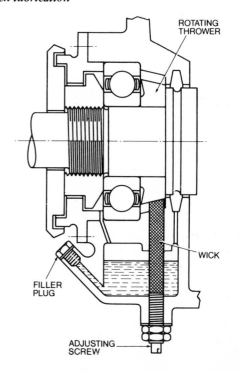

ROTATING THROWER

WICK

FILLER PLUG

ADJUSTING SCREW

This method of applying a small amount of oil to the bearing is sometimes used where high speed, light load, and normal temperature conditions require a minimum amount of lubricant. The oil is fed by the wick from the reservoir to the rotating assembly.

Oil pump feed

Suitable for bearings operating at high speeds, especially where heavy loads generate additional heat within the bearing. The oil lubricates and cools the bearing. The oil is directed at the outside circumference of the inner ring, to reach all working parts of the bearing. It is essential that the oil reservoir formed by the bearing end covers provides lubrication on starting up, until the pump becomes effective. Adequate drainage from the bearing is necessary, since excess oil tends to cause flooding, churning, and overheating.

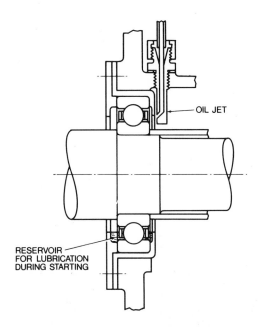

Oil mist lubrication

This system lubricates and cools the bearing, and also helps to prevent foreign matter getting into the bearing housing. It is frequently used on applications such as machine tools, where the same system can lubricate slideways, etc. The design should be in accordance with the recommendations of the lubrication-system or bearing manufacturer, to ensure that all working parts of the bearing are properly lubricated.

The compressed air used must be clean and dry, and the bearings must be continually covered by a thin film of oil. It is advisable to switch on the lubrication system before the bearings start to rotate.

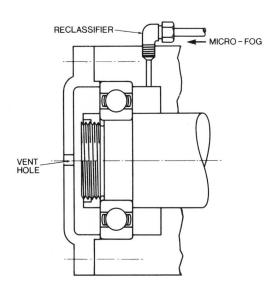

Automatic re-circulating system—vertical shaft

There are several methods of providing a self-contained re-circulating oil system, using rotating flingers or screw-thread devices. A typical system using a rotating flinger is illustrated. It is important to ensure that the oil level when the machine is stationary is below the lip on the bearing end cover.

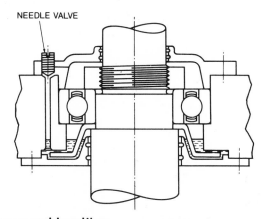

Storage and handling

Storage Bearings are normally treated with preservative oil before packing and despatch. This gives good protection under normal storage conditions. To minimise deterioration in storage, the following points should be observed.

a) Bearings should be stored in their original wrappings. If the wrappings become damaged or contaminated, the bearings should be properly cleaned and protected with a suitable preservative or mineral oil before being re-wrapped.

b) Bearing boxes should be stored flat, and not piled so high that the protective coating is squeezed away from the bearing faces.

c) Bearings should be used in the order in which they are received.

d) The store room should be clean, dry, and correctly ventilated to maintain a normal and even temperature.

Fitting Detailed fitting instructions may be required to ensure that correct assembly methods are used. The following general points should always be considered.

a) A high standard of cleanliness of the bearing, the components to which it is fitted, the assembly area, and the lubricant must be maintained.

b) The shaft and housing seatings must be accurately machined and free from burrs and other damage.

c) When a bearing is an interference fit, pressure must be applied evenly, and only to the bearing ring which is being fitted, not through the rolling elements.

d) Bearings which are an interference fit on the shaft may be heated in clean mineral oil up to 100°C maximum to facilitate fitting. After fitting and cooling, ensure that the bearing is correctly seated against its abutment face.

e) When assembling separable bearings, care must be taken to bring the components together squarely, without damaging the rolling elements or raceways.

Bearing removal Bearings should be removed as seldom as possible. When removal is necessary, a suitable extractor should be used, pulling on the correct bearing ring. Unless such an extractor is used, the bearing may be damaged. If for any reason it is necessary to return a bearing for investigation it should not be cleaned or dismantled.

Bearing cleaning Bearings should be cleaned by immersion in a solvent such as clean white spirit or clean paraffin, then thoroughly dried in a jet of clean, dry compressed air. Bearings should not be spun by the air jet, as skidding can damage the rolling elements and raceways.

Bearing inspection Bearings with visible signs of corrosion, overheating, or damage due to fretting or creep should not be re-used.

Bearings which appear to have been running normally may be inspected for wear or roughness by hand rotation and, if necessary, by measuring radial internal clearance. The rolling elements of separable bearings may also be inspected visually. If all these checks prove satisfactory, the bearings should be thoroughly oiled and protected for storage until they are re-fitted.

Shaft and housing design

Rigidity The shaft must be sufficiently stiff to avoid excessive slope at the bearing under load, and the housing must be sufficiently rigid to avoid distortion of the bearing seating.

Accuracy It is important not to exceed the manufacturer's recommendations for the permissible misalignment due to deflection and tolerance build up. Seatings should also be machined to the manufacturer's specified limits to suit conditions of load and rotation. Errors of form such as taper and ovality should never exceed the shaft or housing tolerance, and should be controlled to finer limits where high speeds or heavy loads are involved, or where a particularly high standard of reliability is required.

Support Ideally the bearing seating should extend over the whole bearing width, and it should not be less than two-thirds of the bearing width. If the bearing outside diameter is used as a spigot for registering end covers or split housings, care should be taken not to distort or overload the bearing ring. If there is any possibility of this, a separate spigot should be used. Abutment faces should always be square to the axis of rotation and provide sufficient axial support for the bearing.

Closures The housing and end covers must protect the bearing from foreign matter, and must be compatible with the lubrication system.

Split housings These must be accurately registered before machining, and preferably designed so that they cannot be fitted incorrectly. Distortion of the bearing outer ring should be avoided.

End covers Where grease lubrication is employed, the recess in the end covers should be restricted to one quarter to one-third of the bearing width. For vertical shafts the space on the upper side of the bearing should be restricted to one-quarter of the bearing width, to minimise the danger of grease slumping.

The lubricant entry point should be diagonally opposite the drain and vent hole, so that fresh lubricant passes through the bearing.

It is preferable to spigot the end covers in the housing bore and not on the bearing outside diameter.

If quiet running is important, the end covers should be designed to avoid resonance.

Ferrous shafts and housings Ferrous materials give good support to the bearing rings, and have a coefficient of linear expansion similar to that of the bearing steel. (12×10^{-6} per °C). The recommendations for standard bearings are suitable for most applications, but special consideration may have to be given to the seating limits for hollow shafts, very heavily loaded bearings, and high-speed applications where there may be a significant loss of gripping pressure.

Non-ferrous shafts and housings Two main problems arise. These are to maintain the correct fit for the bearing rings over the complete working temperature range, and, in light alloy housings, to prevent the bearing seating becoming mis-shapen under heavy and/or shock loading conditions.

It is normal to aim at the recommended seating limits at the working temperature, taking into account the difference in expansion rates of the bearing and seatings; this may produce heavy interference at the opposite extreme of temperature. To allow for this, sufficient clearance must be provided in the bearing. Where the differential expansion is large, and/or where heavy loading is likely to cause housing deformation, the use of a suitable steel liner is recommended. The liner should be an interference fit in the housing at the extreme limits of temperatures.

Chapter 38

Adhesive bonding

Many designers treat the use of adhesives to form load-carrying joints with some mistrust. The basic principle though, is that the joint must be designed from inception for adhesive bonding, and adhesives can be used to form extremely strong durable bonds with metals, glass, rigid plastics, rubber, and many other materials. Bonded joints can give outstanding performances under a wide range of ambient conditions (for example, in environments combining heat with corrosive attack), and designers have for many years used bonded joints where previously riveting and welding was employed. Adhesive bonding also makes possible the production of structures not otherwise feasible because of technical or cost limitations.

The advantages of using adhesives are as follows.
1 Stress distribution over the whole bond area. Local stress concentrations present in riveted or spot-welded joints are avoided.
   The distribution of stresses achieved with adhesive bonding permits reduction in the weight and cost of the structure while improving joint strength and fatigue properties.
2 Lower production costs, due to the elimination of hole drilling and other machining operations.
3 Stiffness increases, despite the weight reduction. Fig. 38.1 shows an example of how a joint may be designed to take advantage of the stiffening effect of bonding. The adhesive forms a continuous bond between the joint surfaces; rivets and spot welds pin the surfaces together only at localised points. Bonded structures are consequently much stiffer, and loading may be increased (by up to 30-100%) before buckling occurs.

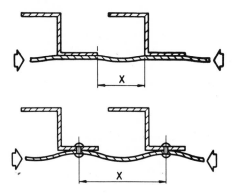

Fig. 38.1   Stiffening effect—bonding and riveting compared. X = unstiffened area

4 Fail-safe properties. In bonded structures, fatigue cracks propagate more slowly than in riveted, machined, or chemical-milled structures, because bonds act as crack stoppers.

5 Gap-filling properties. Certain adhesives are gap-filling, and this makes possible the continuous joining of materials where the gap along the joint is of irregular width.
6 Smooth finish. Bonded surfaces are smooth and unbroken, and are not marred by the irregularities caused by lines of rivets.
7 Joining fragile materials. Metal foil, ceramics, and similarly delicate or brittle materials normally difficult to join are readily bonded.
8 High-strength bonds can be formed at room temperature with minimal pressure by using cold-setting adhesives.
9 The film formed by the adhesive resists corrosion and can form a leak-proof seal and insulate dissimilar metals against electrochemical action.

Against these advantages must be set the following limitations.
1 Surface preparation. Bond strength is dependent on the thoroughness of pre-treatment of the surfaces to be bonded.
2 Usable life. Two-part adhesives have, after mixing, only limited usable life. Bonding on a production scale must be planned accordingly.
3 Support during cure. Some adhesives do not give instant tack. Jigs and fixtures may be needed for assembly, as the joint must be supported during the curing process.
4 Heat stability. Bond strength decreases with increasing temperature. Joints bonded with hot-setting adhesives have the greatest resistance to heat, but even these are unsuitable for structures designed to operate for long periods at temperatures over 150°C.
5 Capital equipment. Initial costs for jigs and ovens for hot-set adhesives may be high when the parts to be bonded are unusually large.
6 High standards of quality-control are required. Partial failures due to poor quality-control are very difficult to detect.
7 Setting times can cause storage and production-control problems.

Terminology associated with the use of adhesives
Structural adhesives. These are load-carrying adhesives where failure would have serious effects.
Non-structural adhesives are adhesives where failure is relatively unimportant.
Thermosetting adhesives show irreversible chemical changes on setting and curing. The subsequent application of heat will ultimately destroy the adhesive. During the setting and curing period, these materials require close temperature-control.
Thermoplastic adhesives are materials which soften with heat. They are applied in the 'hot-melt' form, and the

subsequent application of heat will melt the adhesive without destroying it.

Setting time is the time taken for the adhesive to assume a solid consistency. During the time taken to set, the adhesive may have very little strength.

Curing time is the time required after setting for the joint to attain the designed working strength. The time for curing can vary considerably, from minutes to weeks, depending on the adhesive used.

Solvent-based adhesives. Setting and curing is accomplished by the evaporation of a solvent.

Designing a bonded joint

Joints must be designed for bonding. By following this principle, much better joints will be achieved than if bonding is adopted as a substitute for welding in a joint designed for welding.

Bonded joints perform best under conditions of tension (pure), compression, or shear loading; less well under cleavage; and relatively poorly under peel loading—fig. 38.2.

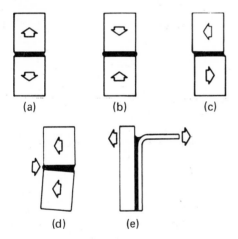

Fig. 38.2   Loading conditions (a) Tension (b) Compression (c) Shear (d) Cleavage (e) Peel

Designing a joint to take pure tensile or compressive stresses is normally impracticable with sheet materials, so all joints in sheet materials should be designed so that the main loading is in shear. Joints between massive parts also perform well in compression or tension loading, provided this is uniform—a side load may set up excessive cleavage stresses in a tension-loaded bond (fig. 38.2(d)).

Where peel stresses may occur, a glass-fabric carrier such as a single layer of open scrim or plain-weave glass-cloth should be incorporated in the joint. Besides improving peel strength, glass-fabric carriers help to retain low-viscosity adhesives in the bond-line during curing, give better

strength retention above the deflection temperature of the cured adhesive, and tend to reduce the effect of stresses due to bonding materials of dissimilar coefficients of expansion. These advantages must be balanced against a slight reduction in room-temperature shear strength and an increase in the cost of the joint.

The simplest-to-make and most commonly used joint in bonded structures is the simple lap shear joint. This has good strength, but, because tensile forces are not in line, the bond area pivots causing stress concentrations under cleavage loading at the ends of the bond (fig. 38.3). Stress

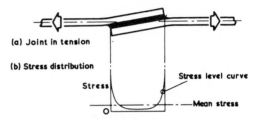

(a) Joint in tension

(b) Stress distribution

Fig. 38.3   Simple lap shear joint

concentrations are reduced and shear strength improved by using a scarfed joint (fig. 38.4), a tapered lap joint (fig. 38.5), a double lap joint (fig. 38.6), or, another version of this, the tubular joint. In practice, tubular joints are tapered (fig. 38.7) so that the adhesive is not pushed out of the assembly.

Fig. 38.4   Scarfed joint

Fig. 38.5   Tapered lap joint

Fig. 38.6   Double lap joint

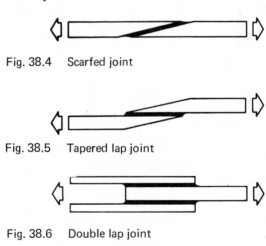

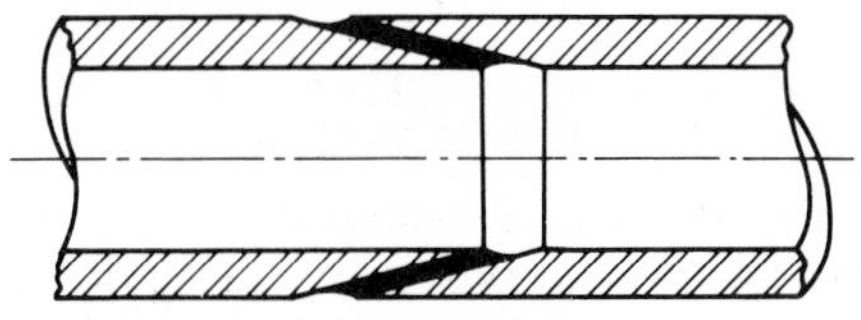

Fig. 38.7   Tapered tubular joint

Tapering the metal reduces stress concentrations due to shear; eliminating eccentricity of loading removes stress concentrations due to bending. These two features are combined in the scarfed joint and the tapered tubular joint. These joints are the most efficient, but the scarfed joint and the modified lap joints mentioned above (tapered and double lap) are rarely used, because they are relatively costly to produce. In practice, the simple lap joint provides the best balance between cost and optimum bond strength.

Bond strength is directly proportional to the width of the joint, and is dependent to a lesser extent on the length of the overlap; doubling the overlap does not necessarily double the strength of the joint. Bond strength is also very largely dependent on the strain in the metal at the ends of the bond. Failure in well-designed bonded joints is initiated by the distortion of the metal which occurs at high strain levels (about 1%).

Bonded-joint design principles applied in practice

The practical use in engineering construction of the design principles given above is shown by the following typical examples.

Simple joints designed to operate under shear. The grinding wheel bonded to its backplate (fig. 38.8) makes particular use of this efficient joining technique, since the stresses set up during operation load the adhesive almost entirely in shear.

Fig. 38.8   Grinding wheel bonded to backplate. The adhesive is exposed to shear stresses.

Double lap joints. The use of double lap joints in the construction of the lamp pillar in fig. 38.9 provides greater robustness while maintaining the saving in weight gained by bonding.

Fig. 38.9   Bonded lamp-post with joints designed to minimise cleavage stresses

Slotted joints. Slotting the joint offers the same advantage as double lapping: elimination of cleavage loading. Typical slotted joints in extrusions, preferably tapered (as indicated) to prevent high stress concentrations, are shown in fig. 38.10

Fig. 38.10   Slotted joints. Tapering removes the high stress concentrations caused by abrupt changes in section.

Reinforcing with fitted doublers. A rod mounted through a plate has only a small area of contact with the plate. To bond the two components, this area is increased either by fitting a collar (fig. 38.11) or by designing the joint according to the standard method for reducing fatigue loading in wheel-to-shaft assemblies (fig. 38.12).

Angle pieces. Two plates to be bonded at right angles are doubled at the contact area with angle pieces. This not only increases bond area, but also greatly reduces the effect of cleavage stresses on the joint. Likewise a right-angle section

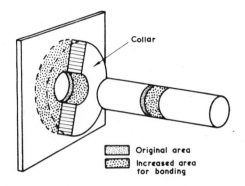

Fig. 38.11 Bond area is greatly increased by fitting a collar

Fig. 38.12 Typical wheel-to-shaft joint design

Fig. 38.13 Angle pieces increase bond area and reduce cleavage stresses

to be bonded to a plate is doubled with an angle piece (fig. 38.13).

Precautions against damage to the joint. Bonded joints give best performance under tension, compression, and shear loading, but performance under peel loading is relatively poor. Joints to be bonded should invariably be designed so that peel stresses are reduced to a minimum. Where there is a possibility of peel stresses being set up by inadvertent misuse or mishandling, it is advisable to reinforce the joint area as shown in fig. 38.14.

Fig. 38.15 shows some simple bonded stiffeners intended to prevent flutter and vibration on thin sheet materials. When the flanges on the stiffening sections deflect with the sheet, little difficulty from peel results. Corrugated backings can provide complete flatness over the entire area.

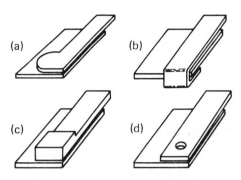

Fig. 38.14 Reinforcement against peel stresses (a) increased bond area (b) Flanging (c) Stiffening (d) Mechanical fixing by spot weld, bolt, etc.

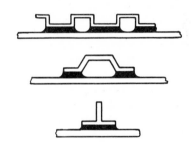

Fig. 38.15 Bonded stiffeners

Peel loading may sometimes be converted into shear loading by the method shown in fig. 38.16.

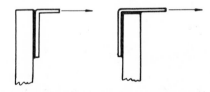

Fig. 38.16 Simple design to convert peeling load into shear load.

Surface preparation and pre-treatments

Epoxy resins adhere firmly to most materials. Bonds of great strength are obtained after removal of grease and loose surface deposits, e.g. rust, from the surfaces to be joined, but, when maximum strength is required, a more thorough mechanical or a chemical pre-treatment is recommended.

Surface preparation by one of the following pre-treatment procedures is essential before bonding:
1 degrease only;
2 degrease, abrade, and again degrease;
3 degrease and chemically pre-treat.

The above procedures are listed in order of effectiveness. Care must be taken to avoid contaminating the pre-treated surfaces prior to bonding. Operators should wear clean white gloves and a surgical cap. Contamination may be caused by finger marking, or by cloths which are not perfectly clean, or by sub-standard degreasing or chemical solutions.

Degreasing. Degreasing is essential even when the surfaces to be bonded appear to be clean. Three methods are given below.

1   Suspend in trichlorethylene vapour.

Or, where a vapour degreasing unit is not available

2   Wipe the joint surfaces with a clean cloth soaked in clean trichlorethylene. Allow to stand for a minute or two to permit complete evaporation from the joint surfaces. Note—trichlorethylene is toxic in both liquid and vapour form. The place of work must be well ventilated with no smoking allowed while vapour remains.

Or,

3   Scrub the joint surfaces in a solution of detergent, e.g. Teepol, or, for metals only, immerse or spray in a suitable alkaline degreasing agent. Wash with clean hot water and allow to dry thoroughly, preferably in a stream of hot air from, for example, a domestic hair-dryer.

Note—the true test of a well-degreased surface is the ease with which it is wetted by the adhesive. If the adhesive film tends to break and gather up into drops, it must be removed and the joint surfaces degreased again more thoroughly.

Abrading. Lightly abraded surfaces give a better key to adhesives than do highly polished surfaces. Abrasion treatment, if carried out, should always be followed by a second degreasing operation; this will also ensure the removal of loose particles. For metal surfaces, remove surface deposits, e.g. tarnish, rust, or mill scale, by grit-blasting. If grit-blasting equipment is not available, or if the metal is too thin to withstand blast-treatment, then clean the joint surfaces with a wire brush, emery cloth, or glass-paper. Repeat the degreasing operation. This will also ensure that loose particles are removed.

Paint surfaces should normally be stripped with a proprietary stripper prior to surface preparation, otherwise the strength of the joint may be limited by comparatively low adhesion of paint to metal.

The treatment described above is sufficient for most adhesive work, but, to obtain maximum strength, reproducibility, and resistance to deterioration, a chemical or electrolytic pre-treatment is required. Care must be taken with the preparation of the chemical solution, not only because of the materials involved, but also because incorrect proportioning may lead to bond strengths inferior to those that would have been obtained if there had been no chemical pre-treatment whatsoever. Time of application is also critical: too short an application does not sufficiently activate the surfaces; whereas overlong application builds up a layer of chemical-reaction products which may interfere with adhesion. For each particular application where chemical pre-treatment is required, the manufacturer of the adhesive should be consulted.

Epoxy resins and hardeners are generally quite harmless to handle, provided that certain precautions normally taken when handling chemicals are observed. The uncured materials must not, for instance, be allowed to come in contact with foodstuffs or food utensils, and measures should also be taken to prevent the uncured materials from coming into contact with the skin, since people with particularly sensitive skin may be affected. The use of barrier creams or rubber gloves is advised. The skin should be thoroughly cleansed at the end of each working period, either by washing with soap and water or by using a resin-removing cream. The use of powerful solvents should be avoided. Disposable paper towels, not cloth towels, should be used. Adequate ventilation of the working area is recommended.

Chapter 39

Production drawings

The following three typical drawings are included as examples of draughtsmanship, layout, dimensioning, and tolerancing.

Figs 39.1 and 39.2 show a shaft and a pulley, and illustrate some aspects of general dimensioning and tolerancing.

Fig. 39.3 shows a partly dimensioned elevation and plan view of a proposed gear-box cover, with a wide application of true-position boxed dimensioning and the associated positional tolerances. To emphasise this style of dimensioning, other dimensions relating to the form of the cover have been omitted.

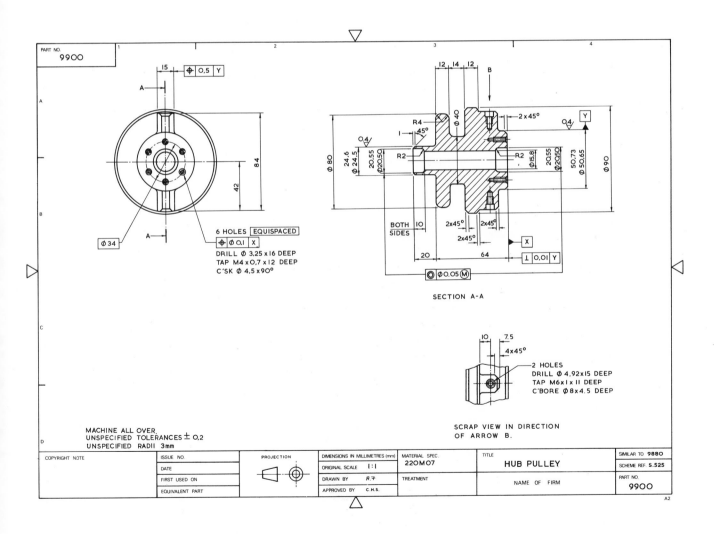

Fig. 39.1

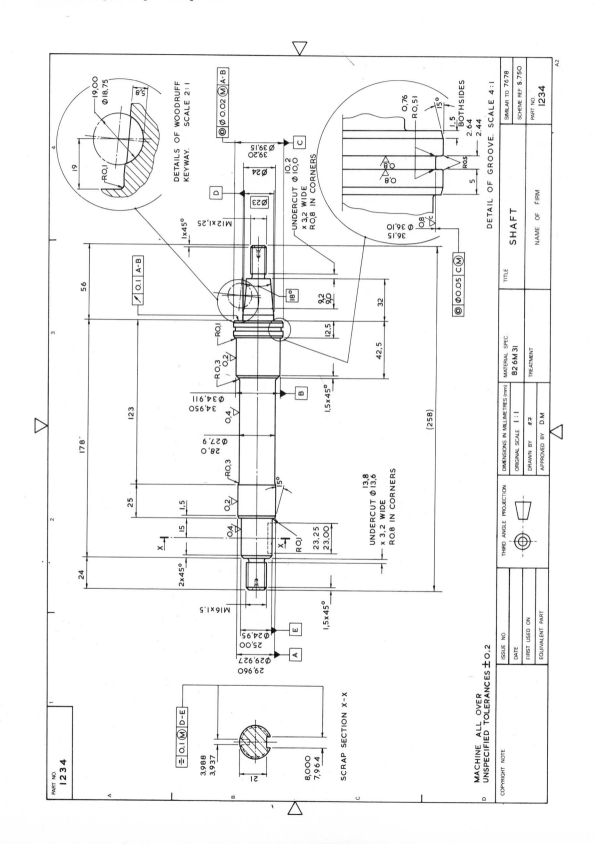

DETAILS OF WOODRUFF
KEYWAY. SCALE 2:1

DETAIL OF GROOVE. SCALE 4:1

SCRAP SECTION X-X

MACHINE ALL OVER
UNSPECIFIED TOLERANCES ±0,2

COPYRIGHT NOTE	ISSUE NO		DIMENSIONS IN MILLIMETRES (mm)	THIRD ANGLE PROJECTION	TITLE	SIMILAR TO 7678
	DATE		MATERIAL SPEC			SCHEME REF S.750
	FIRST USED ON		826M31		SHAFT	PART NO
			ORIGINAL SCALE 1:1			1234
	EQUIVALENT PART		TREATMENT		NAME OF FIRM	
			DRAWN BY R7			
			APPROVED BY D.M			

PART NO.
1234

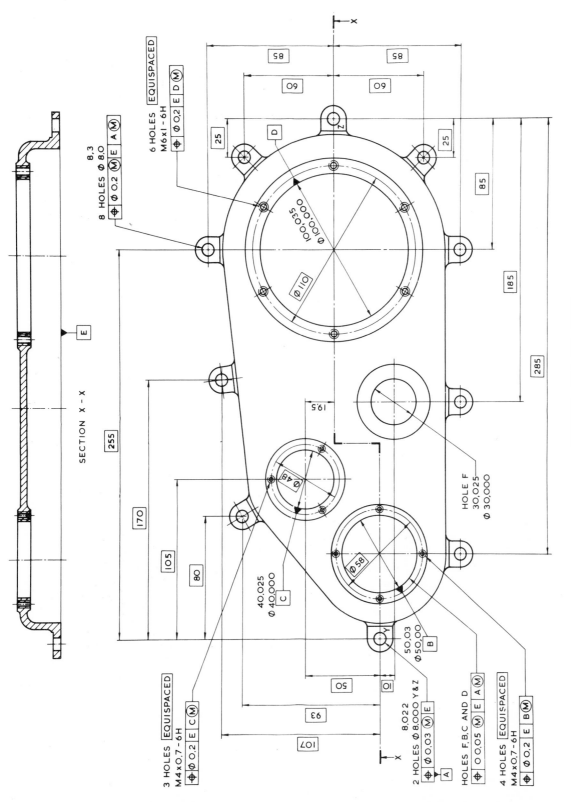

SECTION X - X

PROPOSED GEAR BOX COVER
MATERIAL L.M.4

Fig. 39.3